Remote Control Projects

by

Owen Bishop

BERNARD BABANI (publishing) LTD
THE GRAMPIANS
SHEPHERDS BUSH ROAD
LONDON W6 7NF
ENGLAND

PLEASE NOTE

Although every care has been taken with the production of this book to ensure that any projects, designs, modifications and/or programs etc. contained herein, operate in a corrrect and safe manner and also that any components specified are normally available in Great Britain, the Publishers do not accept responsibility in any way for the failure, including fault in design, of any project, design, modification or program to work correctly or to cause damage to any other equipment that it may be connected to or used in conjunction with, or in respect of any other damage or injury that may be so caused, nor do the Publishers accept responsibility in any way for the failure to obtain specified components.

Notice is also given that if equipment that is still under warranty is modified in any way or used or connected with home-built equipment then that warranty may be void.

First Published – June 1980
Reprinted – April 1985

British Library Cataloguing in Publication Data
Bishop, Owen
Remote control projects (Bernards & Babani Press radio & electronics book; BP73)
1. Electronic control – Amateurs' manuals
2. Remote control – Amateurs' manuals
I. Title
621.381 TK9965

ISBN 0 900162 93 7

Printed and Bound in Great Britain by Cox & Wyman Ltd, Reading

CONTENTS

1 INTRODUCTION

This book is aimed primarily at the electronics enthusiast who wishes to experiment with remote control. Most readers who have already built a few electronics projects should find no difficulty in assembling all of the many projects described here. In addition, many of the designs are suitable for adaptation to the control of other circuits published elsewhere. Full explanations have been given, so that the reader can fully understand how the circuits work and can more easily see how to modify them for other purposes, depending upon personal requirements. Remote control systems have many possible applications, not only in the wide field of model control, but in many other functions in and around the home. Many suggestions are given in the chapters that follow.

The book is also aimed to help the modeller who wishes to go beyond the facilities provided by the ready-built radio-control systems. Other means of transmission (for example ultrasound, or infra-red) may be more suitable than radio in many circumstances and are undoubtedly cheaper. Or the modeller may have particular control requirements that are not met by the systems that are available commercially. Here you will find many suggestions on what can be done by systems that you can design and build yourself. The mechanical aspects of model control are left to the modeller, for it is assumed that he or she is already familiar with the craft. On the other hand, the modeller who is new to electronics will need to learn a few of the skills of electronics construction. With this in mind the illustrations include details of terminal connections of all transistors and integrated circuits used, and there are hints on good construction techniques at appropriate points in the text. There is also an appendix giving essential electronic data. For advice on how to prepare and lay out a circuit board, how to solder, and for many other hints on electronics construction, the reader is referred to *Beginners Guide to Building Electronics Projects*, by R. A. Penfold (Book No. 227 published by Bernard Babani (Publishing) Ltd.).

2 REMOTE CONTROL SYSTEMS

The essential features of a remote control system are shown in Figure 2.1. The controlled device may be a model boat, a domestic TV set, a radio relay station on a remote mountain peak, or a spacecraft on mission beyond Uranus. For each of these controlled devices the system of control has the same features. At one end of the chain of control is the controlled device; at the other end is the operator. The link between them consists of a transmitter, that passes on the commands of the operator to the receiver, by way of the tranmission link. Many types of transmission link are in common use. For controlling spacecraft we must use radio waves and this method is frequently used for model control, but for the domestic TV set we may use either ultra-sound or infra-red radiation. For other purposes we may use visible light, either as a focussed beam or conveyed in an optical fibre. Model electric trains can be controlled by line transmission of electrical signals, sent along the metal track. Since we need to supply the motor of the train with electrical power, it makes sense to use this electrical connection to convey electrical control signals too.

All of the transmission links mentioned above are dealt with in practical detail later but, to understand how control signals may be put into the link and recovered from it we must look more closely at what happens during transmission and reception.

In Figure 2.2 the transmitter and receiver are each shown to be composed of several distinct sections. The transmitter, which is usually considered as a single unit, often hand-held and portable, really consists of 3 sections, each with its own special function. The input interface is usually a keyboard or perhaps one or two joysticks, by means of which the operator communicates instructions to the remote control system. In the simplest system the input interface may be nothing more than a single press-button. If the instructions are to be anything more than 'on–off' or 'stop–go' commands we need

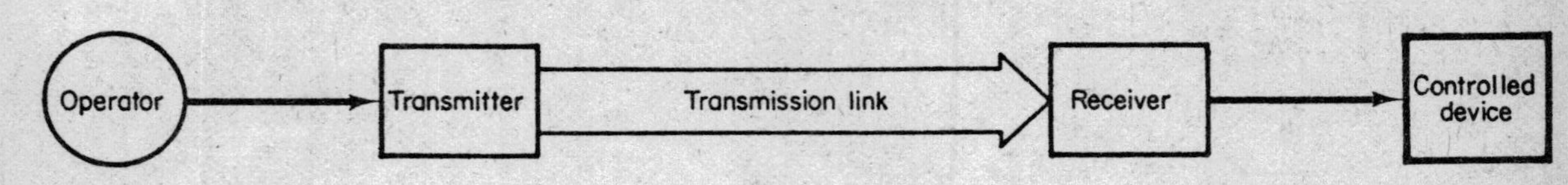

Fig. 2.1 The main features of remote control.

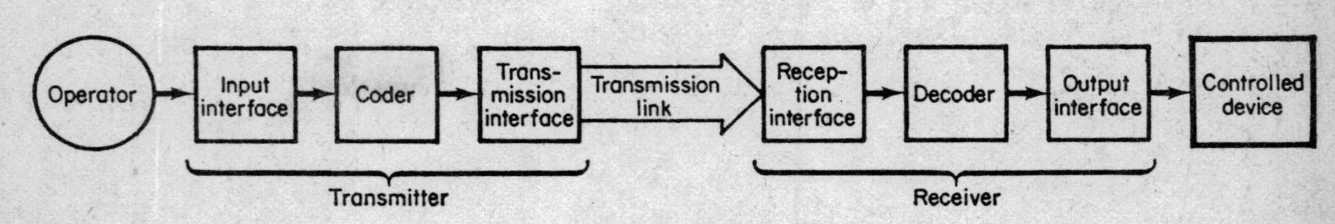

Fig. 2.2 A more detailed view of remote control.

the coder which converts the instructions into a code (usually a series of pulses) ready for transmission. The coded instructions then pass to the transmitter interface which converts them into a form suitable for transmission. For example, the transmission interface may be a crystal specially ground to resonate at 40 kHz, together with an oscillator that operates at this frequency. The coded instructions are passed from the interface as a series of pulses of ultra-sound, and radiate into the surrounding air. Or the interface may be a lamp that emits infra-red radiation. It may be a radio transmitter. For line transmission the interface will be a transistor circuit that injects the coded signal into, for example, the model railway track. A coded signal sent along a single pair of wires can convey quite a lot of information and has several message-carrying applications in the home.

Having looked at the transmitter in this way, we appreciate that whether we are using radio, ultra-sound, infra-red or any other transmission medium we can use the same kind of input interface and coder together with the appropriate kind of transmission interface. Hence the basic circuits described in this book can be combined in dozens of ways to effect many different kinds of remote control.

At the other end of the transmission link is the receiver interface to detect the transmitted commands. The output from this stage is fed to a decoder, that registers these instructions. The same decoder can be used with a radio receiver, an ultrasonic receiver, or a phototransistor (for infra-red or visible radiations). The decoder usually consists of logical circuits that, after registering the coded message, interprets them as control signals to be passed to the controlled device. Between the device and the decoder are usually a number of interface circuits, such as a relay or power transistor to supply electric current to motors or lamps in the controlled device. The electromagnetic ratchet system used in the traditional type of automatic telephone exchange in an example of an output interface. More sophisticated electronic circuits act as interfaces to the circuits of a TV set to control station change, picture qualities, and the volume of sound.

Nowadays one stage of Figure 2.2 may be replaced by a device even more complex in its electronics – a microprocessor system or microcomputer. We no longer need a human operator to control the system. Admittedly there must be a human at work somewhere, preparing the program for the microcomputer, but when in action the whole system can be totally automatic. One could in fact have a microcomputer as the controlled device too. The subject of interfacing to microcomputer is dealt with in a later chapter.

3 SIMPLE ULTRA-SONIC TRANSMITTER

As a practical introduction to remote control technology, this chapter and the next describe a simple system making use of ultra-sound as the transmission link. Ultra-sound is nothing more than sound of a frequency so high that it cannot be detected by the human ear. Figure 3.1 shows the circuit of an oscillator that operates at 40 kHz and is capable of driving a crystal transducer. The transducer (or transmitter, as it is sometimes called) consists of a crystal specially ground so as to resonate vigorously when a 40 kHz signal is applied across its terminals.

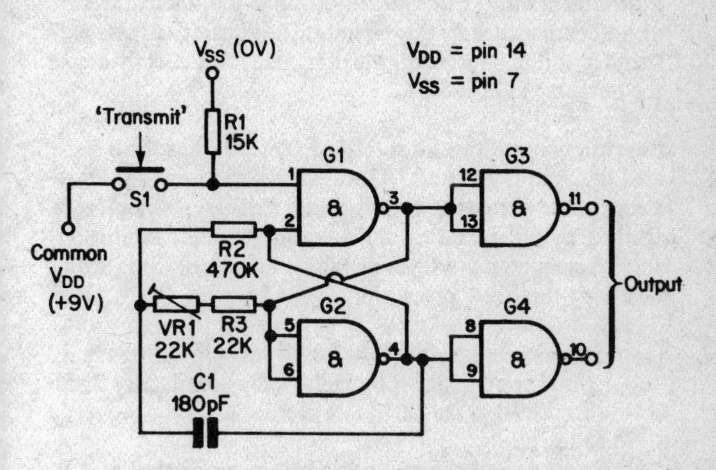

Fig. 3.1 Simple ultra-sonic transmitter oscillator based on CMOS IC (4011)

The oscillator circuit is built from four NAND logic gates and, since the CMOS 4011 i.c. contains four such gates, we need only one i.c. and a few external components. These can all be accommodated on a very small piece of 0.1 inch matrix strip-board. Only a small battery is needed (a PP3), so the whole transmitter can be contained in a small case, making it ideal for a hand-held control.

The heart of the circuit is the astable multivibrator, consisting of two gates (G1, G2) and their associated components. The multivibrator alternates between two states. In one state the output of G1 is high (+9V) and that of G2 is low (0V). In the other state the output of G1 is low and that of G2 is high. The multivibrator changes state at a rate dependent upon the values of R3, VR1 and C1. With the values given in Figure 3.1 change of state occurs 40 000 times a second, giving the 40 kHz output required for making the transducer crystal resonate.

The outputs from the multivibrator are not perfectly square waves, so they are fed to a second pair of gates (G3, G4) to square them off. These gates invert them too, though this is of no consequence in this circuit. The squares outputs are then fed to the transducer, which is connected directly across the outputs of G3 and G4.

Construction presents no special problems except those associated with all CMOS i.c.s. These are prone to damage by electrostatic charges and, though the manufacturers protect the i.c.s by incorporating diodes to short-circuit externally applied charges, it is wise to eliminate the risk of damage by observing these few precautions when handling CMOS i.c.s:

(1) Suppliers usually send the i.c. to you with its terminal pins shorted together by metal foil or conductive foam (black); leave the i.c. in this packing until you are ready to use it.
(2) Carry out all construction work on an earthed metal surface. The author uses an inverted lid from an old biscuit 'tin'.
(3) Do not wear clothes made from synthetic fibre (nylon, terylene, etc.) when handling CMOS i.c.s for these are liable to generate electrostatic charges. Preferably roll up your sleeves and rest your wrist or forearm on the metal sheet when handling the i.c.s.
(4) Earth the bit of the soldering iron.
(5) Touch other metal tools (wire-strippers, screwdrivers, etc.) against the metal sheet to discharge them

immediately before use.

(6) Build the circuit without i.c.s, then solder the i.c.s last.

(7) When testing partly-built circuits, the power supply to the i.c. must be on before high inputs are applied to other pins. When testing is completed, the power supply should be disconnected last.

One way of minimising danger to the i.c. is to mount it in a socket. The circuit is built first, including the socket for the. i.c. When all is complete and checked, the i.c. can be inserted in the socket, observing precautions 1, 2 and 3 as listed above. Using sockets simplifies the procedure and makes it much easier to remove the i.c. later should this be necessary. On the other hand, the price of the socket may be greater than that of the i.c., so direct soldering and careful observance of all the precautions can save money. The use of Soldercon pins instead of conventional i.c. sockets is a satisfactory compromise for those who do not wish to risk soldering the i.c.s directly to the circuit board. The beginner should not be put off by the grim warnings above. Using the above procedure, and even with occasional careless lapses, the author has never damaged any i.c. among the many hundreds handled.

Testing the circuit

Before the transducer is connected the oscillator circuit can be tested by using an oscilloscope, if available. Alternatively the action of the circuit can be slowed down by connecting a high-value capacitor in parallel with C1. A capacitor of, say, 100μF, gives a frequency of about 0.1 Hz (one oscillation in 10 seconds) giving you plenty of time to check the outputs of the gates with an ordinary voltmeter. The frequency of oscillation is finally set to 40 kHz by adjusting VR1, but this is best done when the receiver has been constructed. The transmitter and receiver may then be adjusted to obtain maximum operating range.

Using the transmitter

This circuit has no coder (see Chapter 2) so only on–off or stop–go commands may be transmitted. This is sufficient for many applications, such as stopping or starting an electric train, or switching a radio set on and off. In Chapter 5 there are circuits that improve upon this by allowing the controlled device to step through a sequence of different operations each time a pulse is received. In other applications a coded set of pulses may be required and for this purpose a coder may be connected to this transmitter circuit (Chapter 8).

4 SIMPLE ULTRA-SONIC RECEIVER

The circuit of a simple, yet very effective, receiver is shown in Figure 4.1. The ultra-sound is picked up by the ultra-sonic receiver crystal, RX1. This is prepared during manufacture so as to resonate strongly to ultra-sound of frequency 40 kHz. Thus it resonates strongly when it detects a signal from the transmitter described in Chapter 3, but is virtually unaffected by sounds of other frequencies. This gives freedom from spurious triggering of the circuit. The electrical output from the crystal is amplified by TR1 and TR2, rectified by D1, and produces a drop in the potential difference across R6 when a signal is being received. The operational amplifier, IC1, is affected by the consequent reduction in the current flowing to its inverting input (pin 2), and its output voltage (pin 6) rises. This raises the potential of both plates of C3, causing an increase of potential at the non-inverting input (pin 3), latching the i.c. to give continued high output. Eventually the additional charge on C3 is discharged through R9, and the i.c. becomes unlatched, (assuming that the ultra-sonic signal has ceased in the meantime) allowing the output to fall again. The variable resistor VR1 is used to set the level at which triggering of the circuit into latching condition occurs. If the value of C3 is increased to, say, 150μF, the latching action is prolonged for a period of about 10 seconds. This could be made use of if it was required that a short ultra-sonic signal should evoke a prolonged response.

The output of IC1 is also fed to the potential-divider, R10/R11, so that TR3 is switched on when the output of the i.c. goes high. An LED can be used as shown to indicate the state of the circuit, the LED coming on when a signal has been received and going off a fraction of a second after the signal has ceased. The output of the circuit is taken from the collector of TR1, and is high (about 10.5V) when no signal is being received, dropping sharply to low (less than 0.1V) when a signal is detected. The values quoted apply to operation from a 12V supply, but it may be more convenient to operate the receiver from a battery giving 9V or 6V. Though this is lower

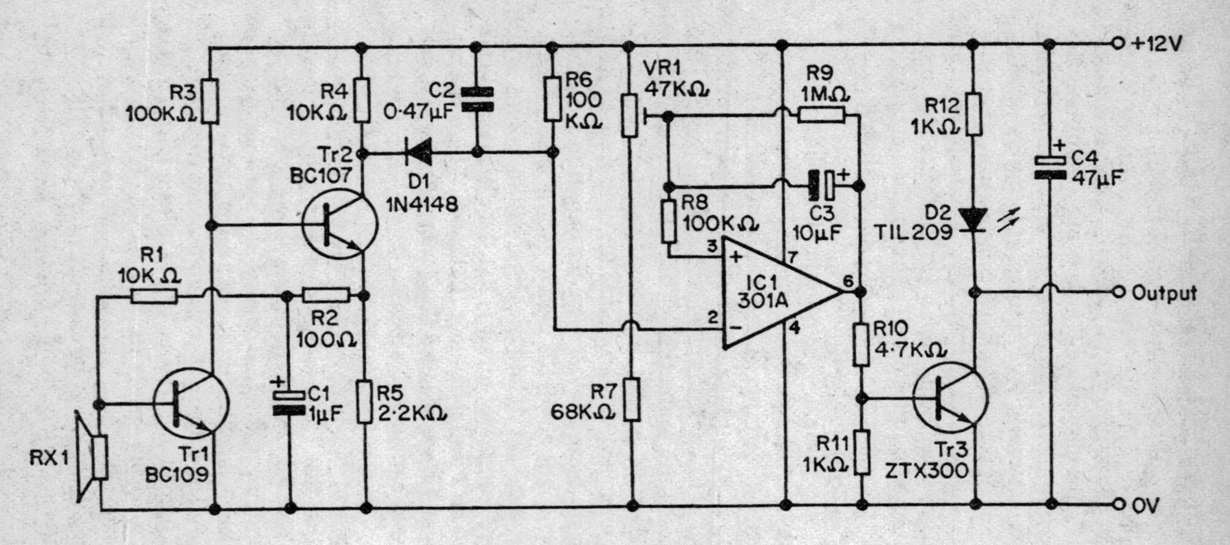

Fig. 4.1 Ultra-sonic receiver.

than the voltage recommended for the i.c., this seems to make little difference to sensitivity and no changes in component values are required (though VR1 will need a different setting). If a 6V supply is used, the output voltage is 5V when high, and close to 0V when low, making the receiver completely compatible with TTL circuits. At 6V the receiver requires only 12mA, even when the LED is lit, thus making this circuit very suitable for battery operation in model control projects. Whether its output is analysed by TTL or, to obtain maximum power economy, by CMOS, depends on the kind of application. If CMOS is to be used, the PP3 battery is a conveniently small power supply, and both the receiver and CMOS i.c.s can operate on the 9V it supplies.

Construction

With reasonably compact layout the whole circuit can be accommodated on a piece of 0.1 inch matrix strip-board, about 15 strips wide and 30 holes long. This has room for an i.c. to analyse the output; if a more complex decoding circuit is required, the board should be larger or the decoder placed on a separate board. The receiver circuit presents no problems in assembly. The ultra-sonic receiver RX1 has two terminals, one of which is connected to its casing; this terminal should be connected to the ground line of the circuit. The whole circuit should be completed and checked for correct wiring, and the absence of solder bridges and other construction faults before it is tested. If the LED has been incorporated, this can be used to see if the circuit is working properly. If it has been decided not to include the LED, connect a voltmeter to the ground line and to the output terminal. When power is first applied the LED normally emits a single flash, though this may not happen at first testing. Instead the LED may shine continuously, indicating that the wiper of VR1 is set too near to the positive rail. If this is so, turn VR1 until the LED goes off; alternatively, if you are using a voltmeter across the output, turn VR1 until the voltmeter reads 'high'. The action of the receiver may then be tested by using the ultra-sonic transmitter (Chapter 3) held with its transmitter crystal pointing toward the receiver crystal and about 1 metre away from it. Immediately the 'transmit' button is pressed, the LED should light, and stay lit until a fraction of a second after the button has been released.

If the receiver shows no response, the fault probably lies with the transmitter, since its oscillator circuit (Figure 3.1) may not be in perfect resonance with the transmitter crystal at exactly 40 kHz. Adjust VR1 *of the transmitter* while pressing the 'transmit' button, until the receiver shows response by the lighting of the LED (or output voltage falls to zero). Then gradually increase the separation between transmitter and receiver, keeping the two crystals pointing at one another. Their effect is very directional and, though the system will work well even when they are not at all closely lined up, maximum range can be attained only when they are directed along the same axis. A range of 4m or more should be readily attainable, but much depends upon the nature of the surfaces of walls, floor and furniture. Range is relatively great in a narrow corridor, especially if it lacks carpets and curtains.

When you appear to have exceeded maximum range and the receiver no longer responds to transmissions, try adjusting VR1 of the transmitter to improve resonance. This may give even further extension of range. Try also to improve the sensitivity of the receiver by turning the wiper of VR1 (of the receiver) toward the positive rail until you reach a position in which the LED does not light, but from which position the slightest further movement toward the positive rail causes the LED to light. This is the position of maximum sensitivity.

Properties of ultra-sound

We have defined ultra-sound as sound of very high frequency, inaudible to the human ear. It travels in air at the same speed as audible sound (about 340m/s at normal temperatures) so it follows that ultra-sound has very short wavelength. The wavelength of ultra-sound at 40 kHz is a little less than 8mm. This is far less than the wavelength of the sounds contained in the human voice, which average 2 metres or more. Most of the sounds of everyday life have wavelengths well over 1 metre.

Since the furniture and other objects in a room (including people) have dimensions that are generally smaller than the wavelengths of everyday sounds, the sound waves are diffracted around such objects when they meet them. The objects do not cast any 'sound shadow'. If someone stands between you and the radio, you can still hear the broadcast well; if you are hiding behind a tree you can be heard when you sneeze. Similarly, sounds are *diffracted* when they come to an opening that is smaller than their wavelength. Sounds made inside a room pass through an open doorway or window and can be heard outside. You do not need to be able to *see* a person who is talking in a room to be able to *hear* that person quite plainly.

Ultra-sound is also diffracted by objects or openings as small as or smaller than its wavelength. Since people, furniture and many other objects have dimensions *far greater* than 8mm, they cast distinct *sound shadows*. If a person stands between the ultra-sonic transmitter and receiver, the transmission is blocked. Even a hand placed between them can prevent the transmission. It is important to remember this when using ultra-sound for remote control. Though you may be able to hear the sounds made by the model, the receiver in the model may not be able to hear the ultra-sound from the transmitter. If you wish to transmit ultra-sound from room to room, you must have an open 'line-of-sight' between transmitter and receiver. An alternative is to place a smooth reflective surface at the doorway to catch and redirect the beam from the transmitter.

This property of ultra-sound has a useful application, not connected with remote control, but worth mentioning since it makes use of the same transmitter and receiver circuits. If transmitter and receiver are placed on opposite sides of a corridor or doorway, with the transmitter switched on to transmit continuously, the transmission is interrupted whenever a person walks between them. This can become the basis of an alarm system, or can be used for counting persons or objects passing a given spot in much the same way that a light-beam is used. Being invisible and inaudible, this type of beam is less likely to be detected by an intruder. If the doorway or corridor is narrow, causing the person to pass fairly close to the transmitter, the receiver can be placed *on the same side* of the corridor or doorway as the transmitter. When nobody is passing, no ultra-sound is received but, when a person passes close by, ultra-sound is reflected from the body to the receiver and the alarm system is triggered.

5 SEQUENTIAL CONTROL

Having established one way of effecting remote control (Chapters 3 and 4) and before going on to consider other transmission systems, let us see how the arrival of a signal at the receiver may be made to cause the required action to take place. With the ultra-sonic system the indictor LED can be made to light for as long as the 'transmit' button is pressed and to go out very shortly after it is released. This might be all that a control system requires. If the LED is replaced by a relay, or the collector of TR3 (Figure 4.1) is connected to a power transistor, we can switch a motor or other device on for as long as we press the button. More details of how to drive motors and such devices from transistor circuits and logic circuits are given in Chapter 11, but before studying this it is worthwhile looking at some slightly more complex control routines that can be obtained by using only a very few integrated circuits.

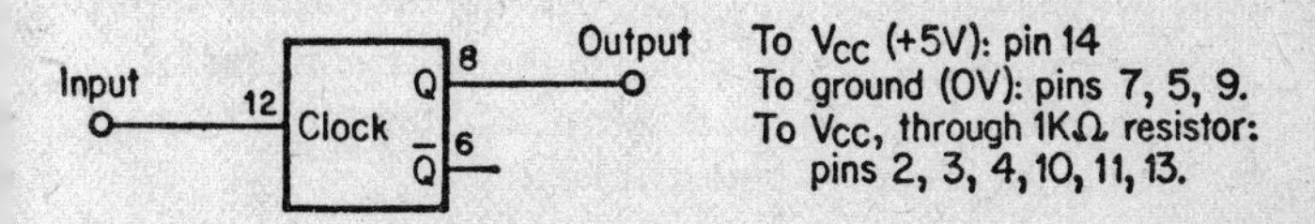

Fig. 5.1 Using a single JK flip-flop, such as the 7470, to provide toggle action.

The simplest one we shall consider is illustrated in Figure 5.1. This has an input, which may be connected to the collector of TR3, Figure 4.1, and an output which may be connected to an LED, a relay, or further logic circuits, as will be described later. The circuitry of the device itself, which is called a J-K flip-flop, is not shown as we are not concerned with the internal details of its working. One input to this flip-flop is called 'clock' and there are other inputs which we are not using in this application. The unused inputs are connected to the positive rail (5V, V_{CC}) through a 1 kilohm resistor, to hold them permanently at 'high' level. Only one resistor is used, one end being connected to the 5V rail, the other end being connected to all the pins listed in Figure 5.1. The flip-flop has two outputs, Q and $\overline{Q}$: if Q is high $\overline{Q}$ is low: if Q is low, $\overline{Q}$ is high. When the clock input changes from high to low, Q and $\overline{Q}$ change state. When clock input changes from low to high, the outputs remain unchanged. Thus if the input is connected to the receiver at TR3 collector terminal and an LED is connected to the Q output of the flip-flop, the state of the LED (lit or not lit) changes whenever the transmit button is pressed. By this simple addition to the circuit of the receiver we have arranged for the LED to go on when the button is pressed and *to stay on*, for hours if need be, until the button is pressed a second time. Here we have the simplest sequence of control: push-on, push-off. This is called *toggle action.*

The next step is to connect two flip-flops in series (Figure 5.2). We can use two 7470 i.c.s, each with its single flip-flop or, for economy in cost, space and wiring, use the 7473 which contains two identical flip-flops. The Q output of flip-flop 1 is fed to the clock input of flip-flop 2. As before, outputs change state whenever clock inputs go low. The effect of this is to produce the sequence shown in the table in Figure 5.2. There are four stages to this sequence, and the outputs change each time the 'transmit' button is pressed.

Electronic circuits of this type are very similar in function to the mechanical arrangements formerly used for remote control, and still used in cheap radio-controlled toys. In these there is some kind of ratchet mechanism which is actuated each time a radio pulse is received. At the mechanism turns, step by step, it brings various gear-wheels or levers into action, so the model goes through a sequence of actions. For example, a toy 'robot' might be made to move forward, turn left, turn right, stop, in sequence. If it is moving forward and you want it to turn right, you press the button twice in rapid succession, so that it skips past the 'turn left' stage very quickly. Similar escapement mechanisms driven by twisted rubber were formerly widely used in model aeroplanes. We can use the same approach with the flip-flop circuit of Figure 5.2, so that when controlling two motors we can have neither, one, the

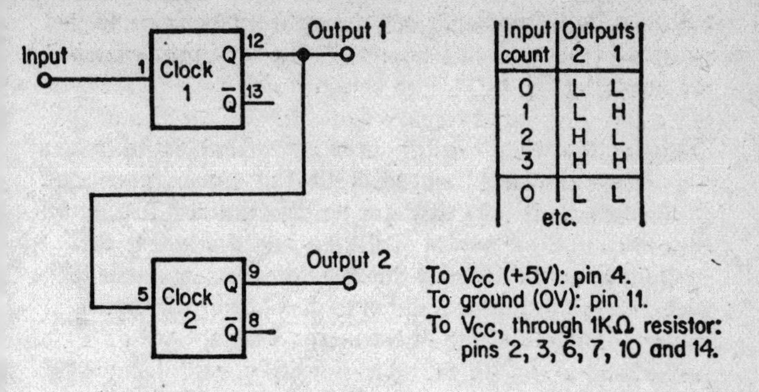

Input count	Outputs 2	1
0	L	L
1	L	H
2	H	L
3	H	H
0	L	L
etc.		

Fig. 5.2 Sequential controller, using a 7473 dual JK flip-flop.

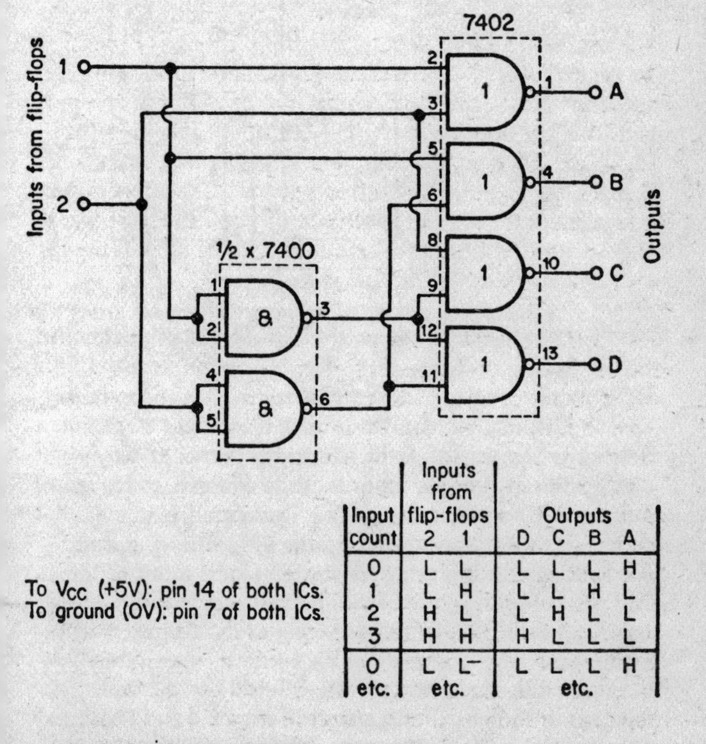

Input count	Inputs from flip-flops 2	1	Outputs D	C	B	A
0	L	L	L	L	L	H
1	L	H	L	L	H	L
2	H	L	L	H	L	L
3	H	H	H	L	L	L
0	L	L	L	L	L	H
etc.	etc.		etc.			

Fig. 5.3 Decoder to activate 1 of 4 outputs in sequence.

other, or both running, and can skip from one state to the other by pressing the button rapidly the appropriate number of times.

The arrangement of Figure 5.2 can control two devices with a 'none-one-other-both' sequence. In this sequence we can recognise four distinct stages, so we have the basis for controlling four separate actions. All that is needed is that the outputs from the two flip-flops be *decoded*. A circuit for this is shown in Figure 5.3. For each of the four combinations of inputs, *one* output channel goes high.

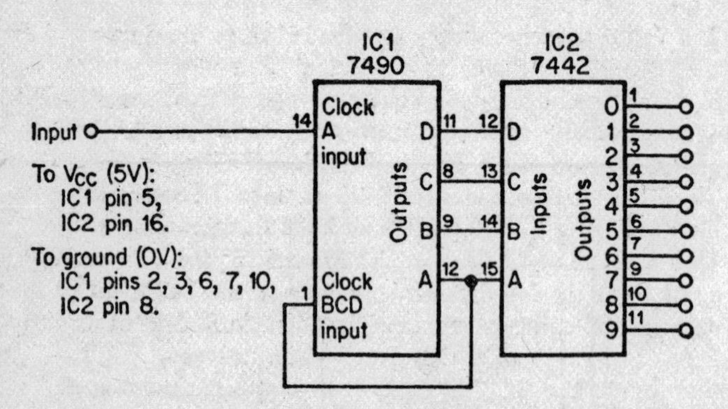

Fig. 5.4 10-stage sequential controller and decoder, using the 7490 decade counter and the 7442 BCD-to-decimal decoder.

If we require a longer sequence, it is best to use flip-flops already interconnected in the form of a counter i.c. Figure 5.4 shows one way of doing this, using the 7490 decimal divider. Each time its input is made low the count at outputs A–D advances one step. The output is in binary form. To decode this we employ the 7442 BCD-to-decimal decoder. As successive pulses are fed to the input of the 7490, the outputs 0 to 9 of the 7442 go high in turn. A ten-stage sequence is likely to be as long as most projects will require, for it is generally inconvenient to have to step through half-a-dozen or more stages quickly to get from one function to another. For sequences between 4 and 10 stages long we can use the 7490 or other dividers, as listed below:

No. stages in seq.	Divider i.c. used	Input to pin	Join pins	NAND outputs from pins	Output from pins
5	7490	1	–	–	– 11 8 9
6	7492	14	1–12	–	– 9 11 12
7	7490	14	1–12	8.9.12	– 8 9 12
8	7493	1	–	–	– 11 8 9
9	7490	14	1–12	11.12	11 8 9 12

The 5-stage sequence simply uses the last three flip-flops of the divide-by-ten chain, the connection from ouptut A to input BCD being omitted. The outputs taken from pins 9, 8, 11 (least significant binary digit first) are fed to pins 15, 14 and 13 of the decoder. Pin 12 of the decoder is grounded. The 6-stage sequence uses the 7492 divide-by-12 counter, but not its D output. Again the outputs are taken to pins 15, 14 and 13 of the decoder. The 7-stage sequence is obtained by using the full divide-by-ten range of the 7490, but detecting when it gets to count 7 by connecting three of its outputs to a 3-input NAND gate (7410). At count 7, all three outputs go high; the NAND gate output therefore goes low; this is wired to the 'reset-0' inputs of the 7490 (pins 2 and 3), which are *not* grounded. The effect is that the counter is immediately reset to zero and begins counting again from

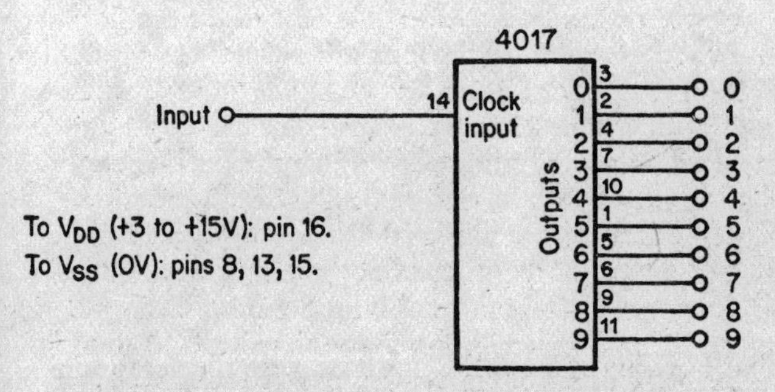

Fig. 5.5 A CMOS version of the circuit of Fig. 5.4, using a single IC, the 4017 counter-decoder.

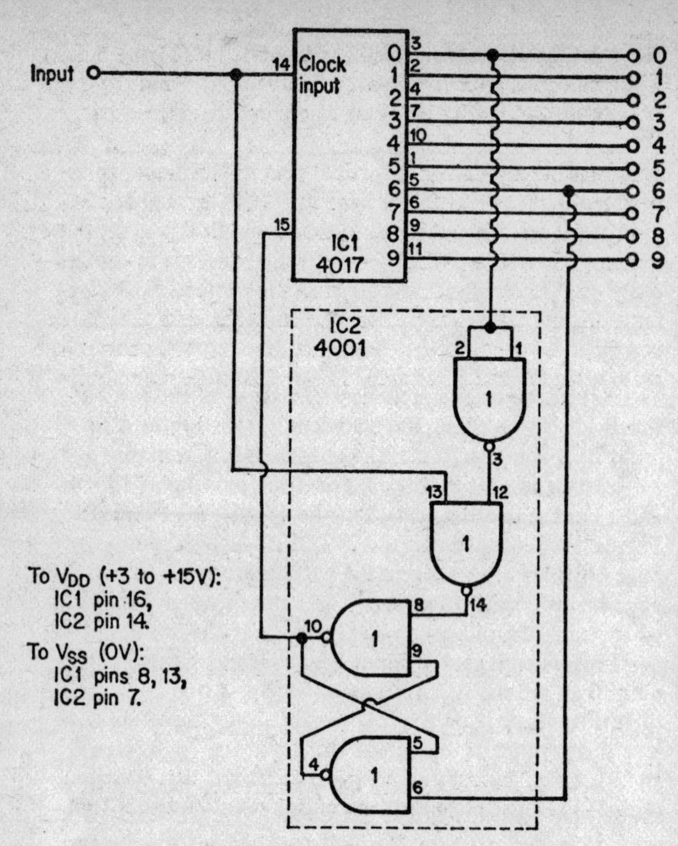

Fig. 5.6 A CMOS counter-decoder connected to reset at the 7th count, making a 6-stage sequence controller.

there. The 9-stage sequence is obtained in a similar way. The 8-stage sequence uses the last three flip-flops of the 7493 divide-by-16 i.c. In this, as well as in the 5-stage circuit, the first flip-flop of the chain is unused, and is available for other purposes if required.

A CMOS version of the above circuits is much simpler to wire up (Figure 5.5) since the divider and decoder are combined in a single i.c. This can be reset at any desired stage by using a set–reset flip-flop constructed from a 4001 i.c. (Figure 5.6). The connections to IC2 are taken from output 0 and from *one* of outputs 1 to 9, depending on when resetting is to occur.

The ways in which these sequential circuits can be put to use is described later but, before leaving this topic, we will briefly look at another type of sequential control. In this system, the sequential circuit is part of the controlled device and continuously steps through each of its stages in sequence. It steps fairly slowly, taking about 2 seconds for each step. Each step corresponds to a different command. The remotely-controlled radio set described in Chapter 11 has a 10-stage sequence:

Stage No.	*Command*	
0	Mute (loudspeaker disconnected)	
1	Full volume	
2	BBC Radio 2	or any other set of pre-tuned stations
3	BBC Radio 3	
4	BBC Radio 4	
5	BBC Radio 1	
6	Manual tuning, medium wave	
7	Record player or microphone	
8	Reduce volume	
9	Reduce volume	

On the panel of the set is an LED 7-segment indicator that shows what stage the circuit has reached at any instant. If you want to switch to Radio 2 watch the changing display and, when '2' appears, press the 'transmit' button of the ultrasonic control. To switch to high volume, wait until '1' appears, then press 'transmit'. To reach the step you want to command may take up to 18 seconds, so this kind of system is of no use for controlling models. But in applications such as the radio set, the delay is no disadvantage. The advantage of this system is that all the complex logic circuits are in the radio set, and are powered from the mains. The remote control transmitter is of the simplest type, as described in Chapter 3; it is small, cheap, its battery lasts for months and it is feasible to have two or more controllers in use.

6 VISIBLE LIGHT SYSTEMS

A big advantage of using visible light as the transmission link is that the transmitter and receiver can be very simply constructed. Indeed, a cheap pocket torch will often act as a transmitter of the single-pulse kind. The great difficulty is that light abounds in the environment, especially by day, so that the transmissions may be subject to considerable interference. One way of reducing the interference problem is to transmit the light in a sharply-focused beam. The light is produced by a small filament-lamp, such as an ordinary torch-bulb, and is brought to a parallel-sided beam by a short-focus lens (Figure 6.1). This is directed at a similar device, with a light-sensitive transistor or phototransistor in place of the bulb. One advantage of this system is that control may be exercised over considerable distances – several tens of metres – but the receiver and transmitter must be visible to each other. For distances exceeding, say, 20 metres, some care must be taken to align the transmitting and receiving devices, and they must be secured firmly against the effects of wind and weather, but once so adjusted, the system is reliable. It has practically no application to model control, except for fixed models, such as model machines, but can have applications for the remote control of many other kinds of electrically-powered device, such as pumps, greenhouse heaters and lights. They are especially useful where it is not permissible or convenient to run wires to carry the transmissions, for example, if a public road passes between transmitter and receiver. An application for the multiple-pulse coder and receiver register is described below.

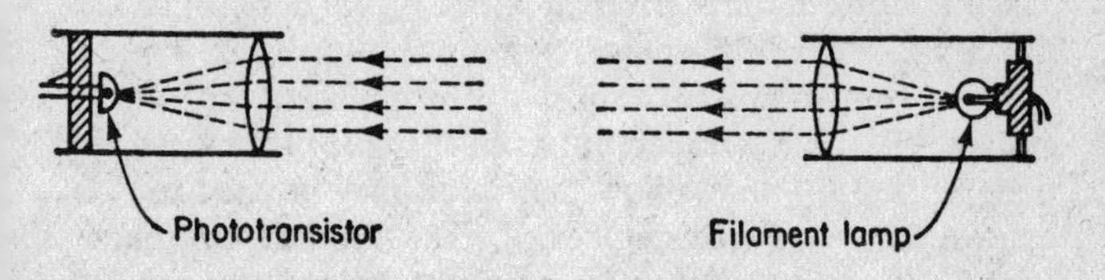

Fig. 6.1 Transmission by focussed light beam.

The multiple-pulse coder transmitter (Chapter 8) can be readily adapted to operate a small filament lamp. The output from the circuit of Figure 8.4 is taken directly to the transmitter circuit of Figure 6.2. The pair of transistors provides gain sufficient to power the lamp which, being slightly overrun, gives a strong signal that will carry for a considerable distance, even without excessively careful alignment of the beam.

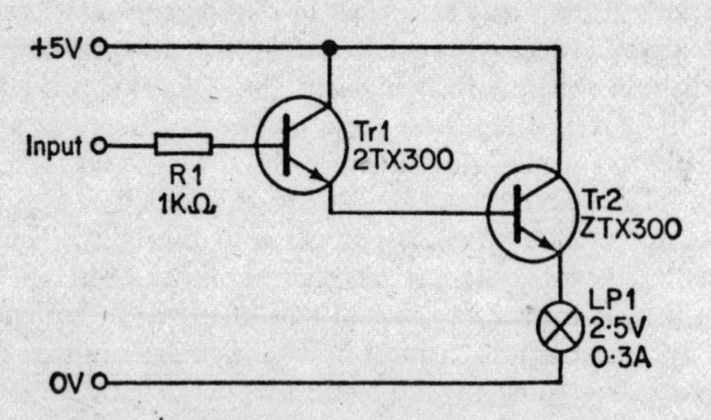

Fig. 6.2 Visible light transmitter interface.

The receiver circuit is equally simple (Figure 6.3). Various types of photo-transistor may be used. One such device is the TIL78, which has no base terminal. When using other types the base connection may be left unconnected as shown in Figure 6.3. If response is weak, a small base current may be provided by connecting the base to the +5V line through a resistor of selected value. For higher sensitivity the photo-darlington transistor type MEL12 is highly recommended. This has a base terminal, which is left disconnected in this application, as in Figure 6.3.

The output from the receiver circuit of Figure 6.3 may be fed directly to the multiple-pulse register circuit of Figure 9.1 and thus the control link is complete. The application of this system is left to the reader, but as an example of what can be done, Figure 6.4 shows how the circuit may be used to control a message receiving device. The output from the latch i.c. is decoded by an i.c. specially designed for the purpose. The 7447 decodes a 4-digit binary number (from

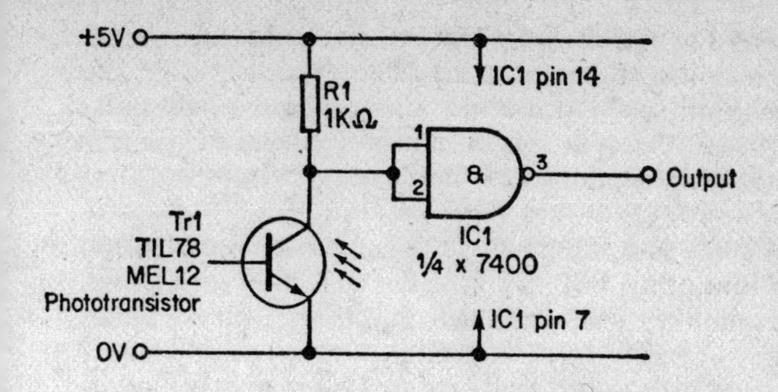

Fig. 6.3 Visible light receiver interface.

decimal 0 to decimal 9) to give the outputs required for driving a 7-segment LED indicator. If a given segment of the display is to be illuminated, the corresponding output of the 7447 goes *low*. Thus to display figure '4', outputs b, c, f and g go low, the remainder staying high; this occurs only when inputs DCBA are low-high-low-low respectively, equivalent to the binary number 0100 (= 4 decimal). Pin 5 of the 7447 is held high permanently so that when all inputs to the i.c. are a low figure zero (0) is displayed. If preferred, pin 5 can be permanently wired to 0V, in which case the display is completely blank when inputs to the 7447 are all

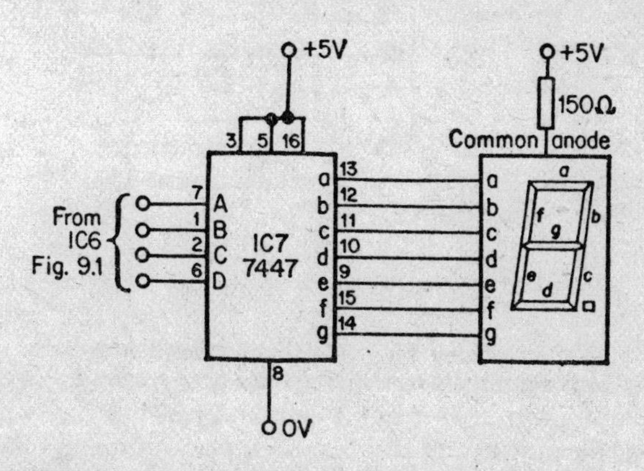

Fig. 6.4 Using a binary-to-7-segment decoder IC to control on LED digital display.

low. For best effect the 150Ω resistor in the 'common anode' connection of the display should be replaced by seven 150Ω resistors, one in each of the wires joining the 7447 to the display. This gives even brightness of illumination, no matter how many segments are switched on.

A device such as this indicates a number corresponding to the setting of the BCD switch of the transmitter (or whatever number-key was pressed on its keyboard). Here is a simple way to send up to 10 different messages on a light-beam. A pre-arranged code enables the person receiving the message to understand which message is associated with each number. As a refinement, an indicator circuit (Figure 6.5) can be added to indicate that a new message is arriving. The simplest and least expensive form of this is to use a 7400 i.c. to provide the three gates required, and to use the decimal point of the LED display to indicate that a new message has arrived. The output from this circuit can, of course, be used to light a separate LED or, with the help of a power transistor, ring a bell or buzzer. Methods for doing this is explained in Chapter 11.

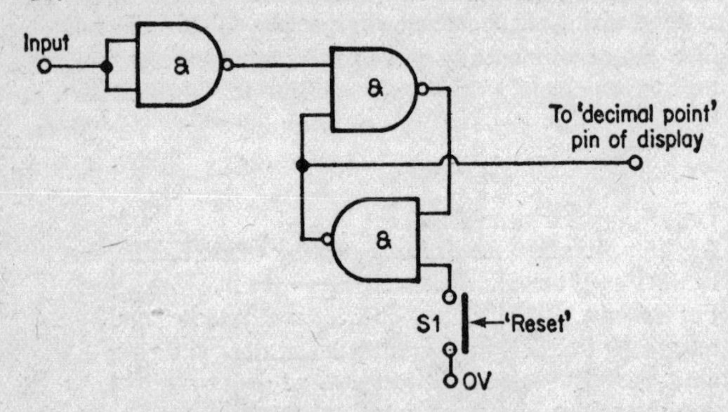

Fig. 6.5 Using the 'decimal point' segment to indicate when a new message has been received.

Optical fibre

As mentioned before, light-beam transmission requires unobstructed line-of-sight between transmitter and receiver, and for long distances alignment is fairly critical. Mirrors could be used for deflecting the beam around corners but this leads to even further alignment problems and there is considerable loss of light on reflection. One way of overcoming these problems is to use an optical fibre. These work by total internal reflection to there is relatively low loss of light during the journey from one end of the fibre to the other. Furthermore, light from external sources can not enter the fibre and cause interference. The transmitting bulb is placed close to one end of the fibre. This junction may be sealed in a dark container to exclude light from other sources. Similarly the phototransistor may be sealed on to the other end of the fibre. Optical fibre may be purchased in lengths up to 100m though as little as ½m may be purchased for experimental use (p. 164). When using optical fibre there are no restrictions about bends and curves in the light path. The fibre is only 1mm in diameter so can be threaded through panels. However, this method of transmission requires physical connection between transmitter and receiver – why not simply use a wired connection? For most purposes a pair of wires is good enough but there can be problems in attempting to interconnect logic systems when the connecting wires are more than 50cm long. Special measures must be adopted at either end of the transmission line and in some instances it could be simpler to use optical fibre. The suggestion is left with the reader as a field for experimentation.

Long distance transmission

At longer distances there is increased risk of changes in local lighting levels being sufficient to trigger the receiving circuit. For example, passing clouds, changes in shadows caused by the relative motion of the sun, dusk, dawn, all produce slow changes of light intensity that can affect the response of the phototransistor. To a certain extent this problem can be reduced or even solved by careful positioning of transmitter bulb and phototransistor. If the transistor is already fully saturated with light, the small amount of additional light from the signal lamp will make no difference to its response. It is therefore essential to shade the phototransistor or its lens housing so that direct sunlight can not fall on it at any time of day. The transmitter bulb should not be placed against a large light-coloured surface, such as white-painted wall because reflection from the wall at certain times of day may swamp the effects of the signal lamp. Even these precautions may be insufficient and the only solution is to use the slightly more elaborate receiving circuit shown in Figure 6.6. The collector of the phototransistor is connected to the inputs of an operational amplifier. The capacitor connected on the inverting-input channel damps voltage swings in that channel. If light intensity changes slowly, the effect of the capacitor is negligible; inputs are more-or-less equal and the output of the amplifier remains close to 0V, equivalent to logical 'low'. The output of the gate is therefore 'high'. However, if there is a *sudden* change in received light intensity, such as is caused by the arrival of a light signal, the imbalance at the two inputs is such as to cause a brief positive swing of the amplifier output. The result is a brief low pulse from the output of the gate. The ending of the light pulse has no significant effect in this circuit. Here we have a way of detecting the arrival of a signal pulse amid a background of slowly changing light levels. The low pulse from the gate (inverted if required) can be used to operate sequential control systems.

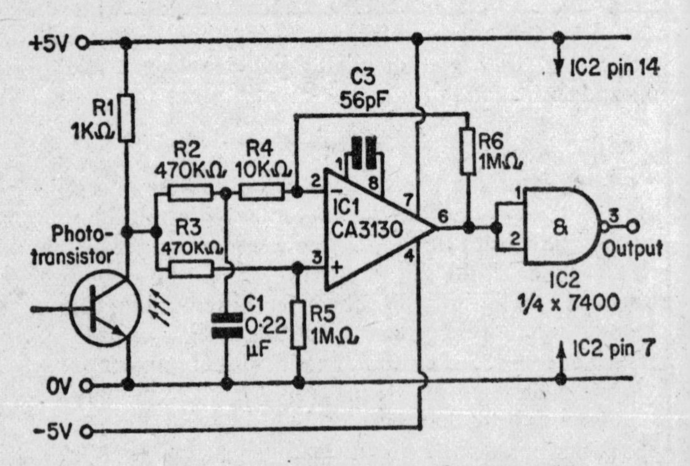

Fig.6.6 Visible light receiver that responds only to rapid changes of intensity.

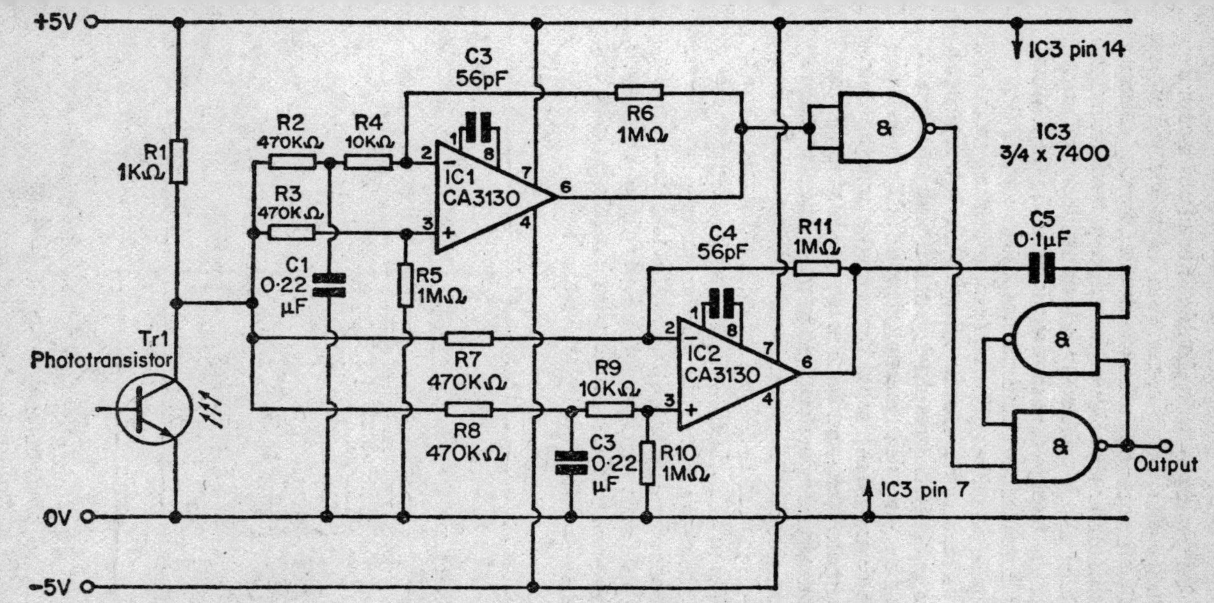

Fig. 6.7 Sensitive light pulse detecting circuit.

The multiple-pulse system (Chapter 8) requires that both the beginning and the end of the pulse should be detected. This is done by the circuit of Figure 6.7. In this circuit we detect the beginning of the pulse in the same way as in Figure 6.6. The end of the pulse is detected by a circuit in which the damping capacitor is wired to the non-inverting channel (IC2). At the arrival of a pulse there is no significant change, but when a pulse ends, the output of the amplifier swings sharply negative of 0V. This negative pulse is transmitted through capacitor to one input of a bistable. The bistable is being used to reconstruct the original signal pulse by being *set* as the pulse begins and by being *reset* as it ends. The output from the bistable has approximately the same length as the original pulse and may then be fed to the register circuit of the multi-pulse system.

7 INFRA-RED SYSTEMS

Infra-red transmission has the advantage over transmission of visible light that it need not be unduly interfered with by external sources of light. In comparison with ultra-sonic transmission, the circuits concerned are simpler and less liable to damage.

A very simple transmitter for single-pulse operation is shown in Figure 7.1a. The TIL38 LED is a large and powerful emitter of infra-red radiation. Its maximum current consumption is 150mA, so the resistor must be chosen to give a current close to that value if maximum range is to be attained. For 5V operation the resistor should be 22Ω, and for 10V operation it should be 47Ω. A single LED without any form of reflector or lens to focus a beam has a range of up to 1 metre. To increase the range one simply adds more LEDs in parallel, as in Figure 7.1b. For 5V operation, the resistor should have the value 5.6Ω; for 10V it should be 12Ω. Three or four such LEDs should be sufficient for control purposes when used in an ordinary living-room or office.

A transmitter of the kind described above can also be driven by a multi-pulse coder as described in Chapter 8, and we will

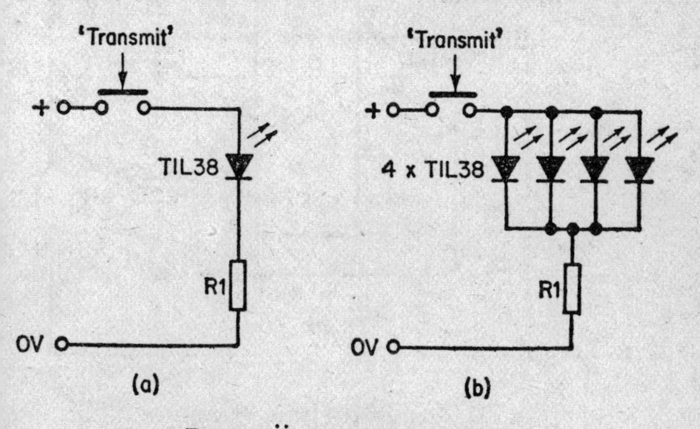

Fig. 7.1 Infra-red transmitters.

later see how infra-red LEDs can be driven from a remote control i.c. of considerable complexity (Chapter 13).

Although the maximum continuous current for the TIL38 is large compared with that of many other types of LED, it can be increased even further provided that this is done for very short periods. If pulses are limited in length to 10μs, and transmitted no more often than once per millisecond, the current through the LED may be as much as 2A, giving a very intense flash of radiation. This is another way of increasing range, though the circuit required is more complex and the use of several LEDs in parallel as in Figure 7.1b is generally to be preferred. A circuit for flashing the LED is shown in Figure 7.2. This uses two CMOS monostable multivibrators, contained in a single i.c. One multivibrator is set to give a 'long flash' (about half a second) on a visible light LED, as an indication that the short flash (10μs) has been transmitted by the infra-red LED. Transmission takes place as the button is first pressed. Though the use of a simple button is adequate for setting-up and testing the circuit, trouble may arise in use owing to contact-bounce, causing several pulses to

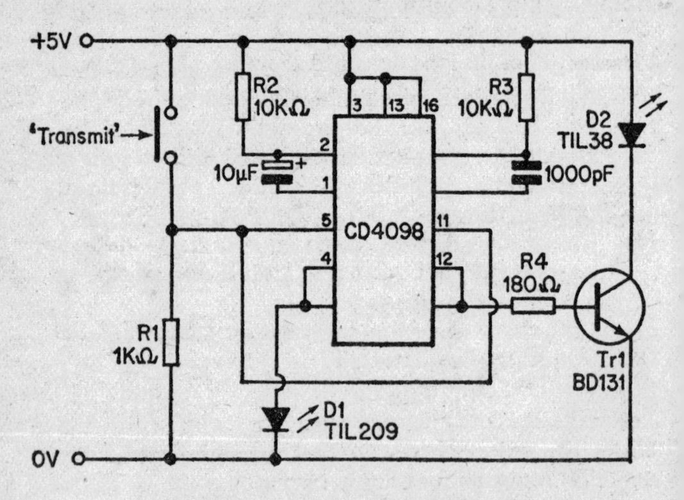

Fig. 7.2 An infra-red pulse transmitter for short light-intensity pulses.

be transmitted. This *may* cause little trouble but, if it does, the triggering of the monostables should be done by a logic input, such as obtainable from a Schmitt trigger gate, or from the output of one of the coding circuits described elsewhere in this book.

One final point concerning transmitters is that the *miniature* infra-red LEDs (e.g. TIL32) are unsuitable for this application because of their low emission. They are intended only for close-range detection, such as in punched-tape reading.

A very simple yet useful receiver is shown in Figure 7.3. It uses a photodiode that is sensitive in the infra-red range. The case of the diode is relatively opaque to visible light, though transparent to infra-red radiation. However, light from domestic filament lamps and high-intensity fluorescent lamps contains a strong component in the infra-red band. So does sunlight. If the circuit is to be used under brightly lit conditions, the action of the infra-red transmitter may be swamped. There are several ways to minimise this:

(1) Screen the photodiode from external sources as far as practicable.
(2) Use a *strong* infra-red source in the transmitter – three, four or possibly more LEDs.
(3) Place a colour-filter over the sensitive surface of the photodiode. Kodak filters 87 or 87C transmit infra-red and absorb most visible wavelengths; the 87C is slightly better in this respect.
(4) Use a circuit that is sensitive only to sharply-defined pulses of infra-red, and not to gradual changes of intensity (see p. 37 later).
(5) Use a circuit tuned to a particular frequency of modulation (see Chapter 12).

The methods above are listed roughly in order of cost and complexity, so it is advisable to try them in the listed order until satisfactory performance is obtained.

The circuit of Figure 7.3 depends upon the fact that the

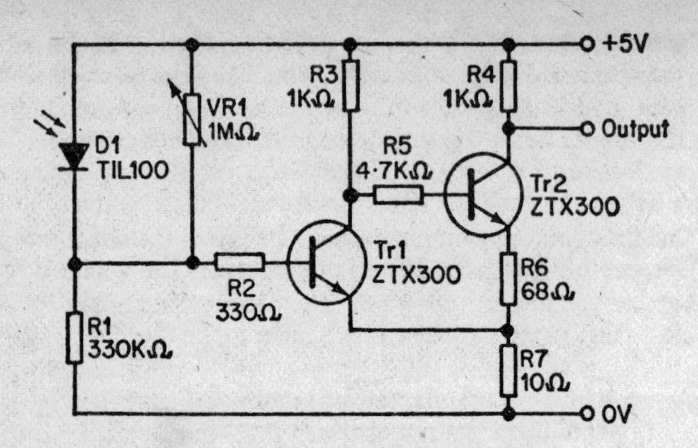

Fig. 7.3 Infra-red receiver interface.

current passing through D1 increases as the amount of radiation falling on it is increased. The result is an increase of potential at the junction of D1 and R1. This causes an increasing base current to flow to TR1, gradually turning it on. As it becomes turned on, the potential at the collector of TR1 begins to fall, so that the base current to TR2 is gradually reduced, turning TR2 off. As TR2 is turned off, the potential at the junction of R6 and R7 falls, for the current flowing through these resistors is being reduced. A fall of potential means that the potential difference between the base and the emitter of TR1 is increased, so TR1 is turned more fully off. This turns TR2 even more fully on. A slight change of potential at the D1/R1 junction produces a rapid 'snap-action', turning TR2 off, and thus producing a 'high' output from the circuit. The level at which this transition occurs can be set by using VR1 to provide a given amount of bias current to TR1. VR1 is set so as to provide *almost* enough current to trigger the circuit. Any *additional* current resulting from a slight increase in the amount of radiation received will be sufficient to trigger the circuit, and cause its output to change from 'low' to 'high'.

Figure 7.4 shows a high-gain amplifier based on two operational amplifier i.c.s. Since the junction between D1 and R1 is coupled to the amplifier by a capacitor, the circuit responds to relatively *rapid changes* of irradiation, such as caused by the arrival of a pulse of infra-red, but not to slower changes such as *gradual changes* in the amount of sunlight reaching the sensor. By reducing the value of C1, this effect can be made even more distinct. The amplifier IC1 is connected as a differential amplifier; the variable resistor is set so that the output of this amplifier, as measured at point A is 7.5V with respect to the 0V rail. This resistor can be of the pre-set type.

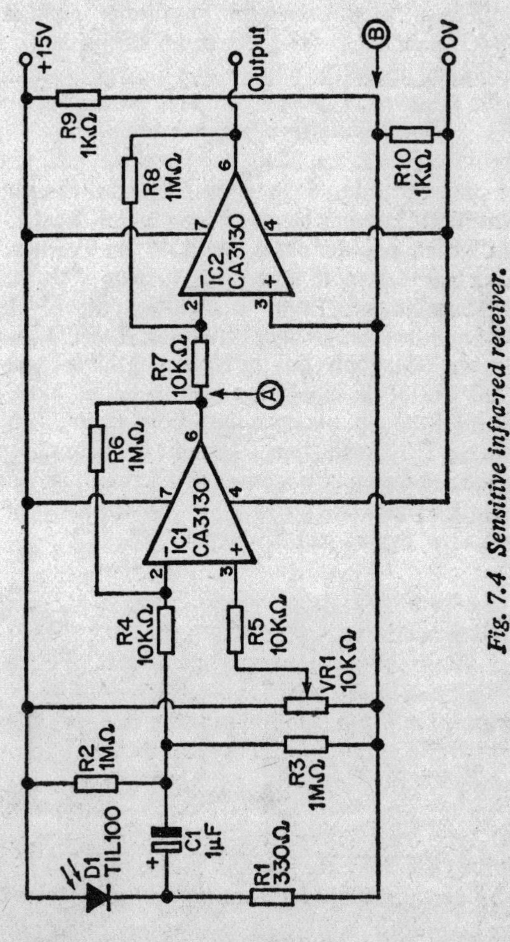

Fig. 7.4 Sensitive infra-red receiver.

It is also possible to operate this circuit without C1. If a wire link replaces C1, the circuit responds to all changes in radiation; the voltage at A goes negative for the duration of a pulse, and rises in between pulses. D1 and R1 may be interchanged if the reverse action is required. In this configuration VR1 may be a potentiometer (volume control) and used to set the level at which the circuit responds.

The single amplifier IC1 may be used on its own if the circuit is found to be sufficiently sensitive. For further amplification, the output from IC1 is fed to a second amplifier, connected as an inverting amplifier. The output from this swings strongly toward the positive rail when an infra-red pulse is received.

The output from these amplifiers, and the circuit next to be described may be used to drive transistors for switching operations (Chapter 11) or may be fed directly to CMOS gates, the CMOS i.c.s being powered from the +15V and 0V lines. Interfacing to TTL is more complicated because of the large voltage swings obtained from the amplifier output. One method of interfacing is given in Figure 7.5. The TTL circuit needs its own +5V supply, but the 0V line of this is connected to the 7.5V rail of the amplifier circuit, indicated by B in Figure 7.4. If the input voltage rises above +5.6V, with respect to the TTL ground line, D1 conducts and so the gate is protected and the input is received as a normal 'high' input. Conversely, if voltage falls below –0.6V, D2 conducts and the input is received as a normal 'low' input.

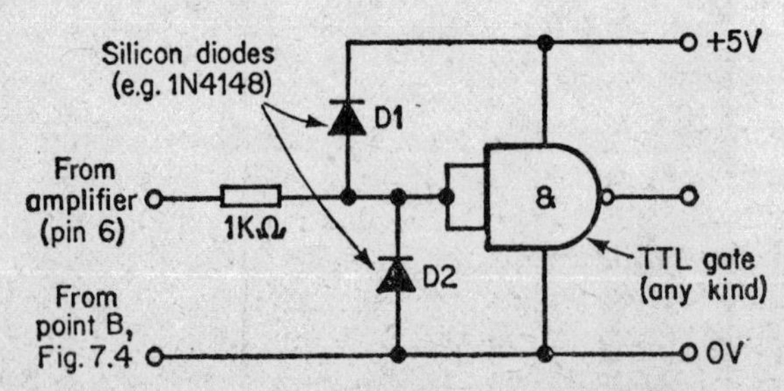

Fig. 7.5 Interfacing the infra-red receiver to TTL circuits.

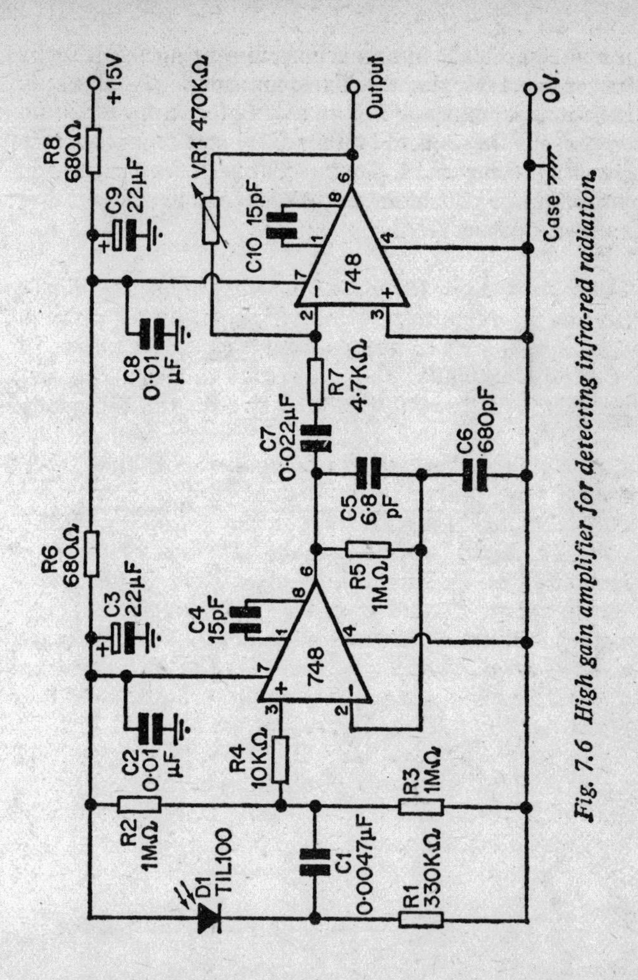

Fig. 7.6 High gain amplifier for detecting infra-red radiation.

An even more sensitive receiving circuit is shown in Figure 7.6. It has two amplifiers, the first being connected as a non-inverting amplifier, and the second as an inverting amplifier. This circuit is coupled to the sensor by a capacitor, so is sensitive to rapid changes of radiation intensity. The second amplifier has variable gain (VR1) allowing for adjustment of sensitivity. One problem associated with the high gain of this circuit is that it is liable to pick up electrical interference from nearby equipment. It should therefore be housed in an earthed metal case, and the ground line connected to the case. The output from this circuit can be used for driving transistor switches, CMOS gates, and the special remote control i.c. to be described in Chapter 14. To drive TTL gates the maximum output voltage must be limited to +5V. An interface circuit is shown in Figure 7.7.

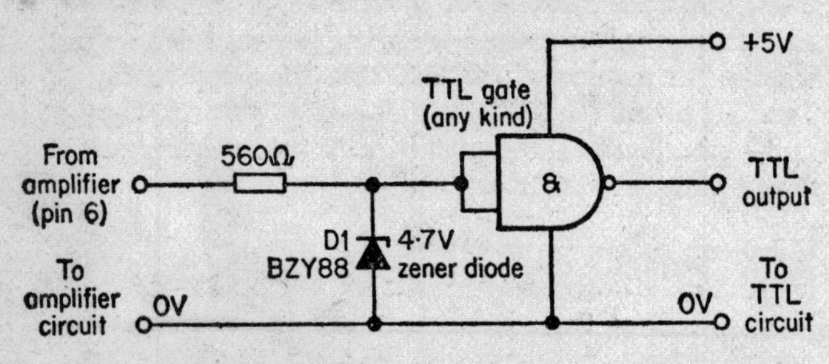

Fig. 7.7 Interfacing the high-gain amplifier to TTL circuits.

8 MULTIPLE PULSE TRANSMITTER

The variety of commands that may be conveyed with a single pulse transmitter is very limited. Although one can increase the scope by stepping through a sequence (Chapter 5) the use of long sequences means delayed response times, and may be unacceptable for other reasons in certain applications. If we use a train of pulses made up according to given rules, we can increase the number of available commands appreciably. There are several ways of doing this, one of which is described in this chapter, and another which is the basis of the system described in Chapters 13 and 14.

Consider a code made up of a train of 4 pulses, of equal length, each of which can be either high (= 1) or low (= 0). We can make up 16 such trains, arranging the pulses in all possible ways, corresponding to the binary numbers from zero (0000) to 15 (1111). This means that we can transmit up to 16 different instructions by means of this code. For example, we could make up a code for controlling a model aeroplane, and this might have the following form:

Decimal number	*Binary number DCBA*	*Corresponding command*
0	0000	No change
1	0001	Centralise control surfaces
2	0010	Move rudder left
3	0011	Move rudder right
4	0100	Move elevators up
5	0101	Move elevators down
6	0110	Move ailerons to roll left
7	0111	Move ailerons to roll right
8	1000	Engine speed 1 (slowest)
9	1001	Engine speed 2
10	1010	Engine speed 3
11	1011	Engine speed 4
12	1100	Engine speed 5
13	1101	Cut engine
14	1110	Landing lights on
15	1111	Landing lights off

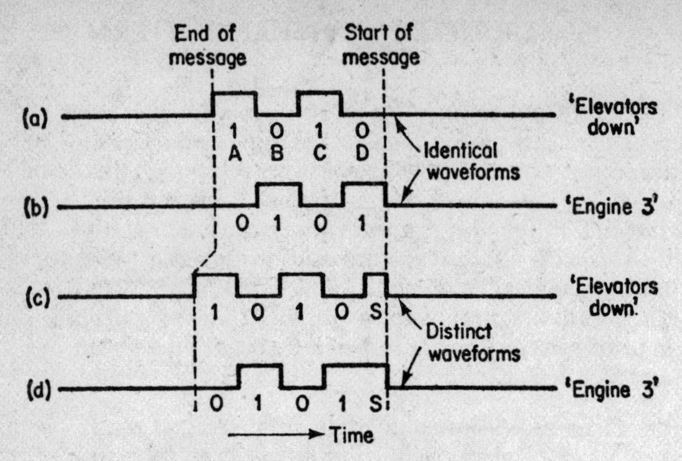

Fig. 8.1 Coded commands.

If we wish the elevators to move down we transmit the code 0101. Figure 8.1a shows the transmitter output, assuming the digits of the code are transmitted in order D, C, B, A. At the receiver this code is indistinguishable from 1010 (Figure 8.1b) engine speed 3, so some means of indicating the start of the command message is needed. One way of doing this is to begin *every* command with a high pulse, followed by the four code pulses. Then we have the messages shown in Figures 8.1c and 8.1d; the first or start pulse (S) is half the length of the other pulses because this allows the circuit to be simpler without loss of effectiveness. To follow the operation of the circuit in more detail look at the block diagram (Figure 8.2).

Pulse length is determined by the clock which produces a continuous series of pulses at approximately 1.7Hz. The system could work just as well with a much higher clock frequency, but the slow clock rate makes it much easier to test the working of the circuit. The clock pulses control the operation of a shift register. This contains a chain of 5 registers, each of which holds data. A 'set' register holds data '1' (output high); a 'reset' register holds data '0' (output low). As the clock output goes from low to high, data is shifted from each register to the next register along the chain. In

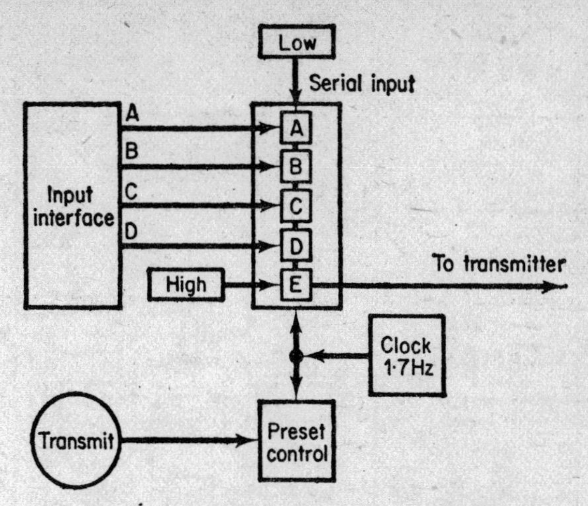

Fig. 8.2 Block diagram of coder circuit.

Figure 8.3 we see what happens. To begin with all registers are 'reset', with low outputs. When the operator presses a command key the coded command data appear on the lines A, B, C and D from the input interface. These can be highs

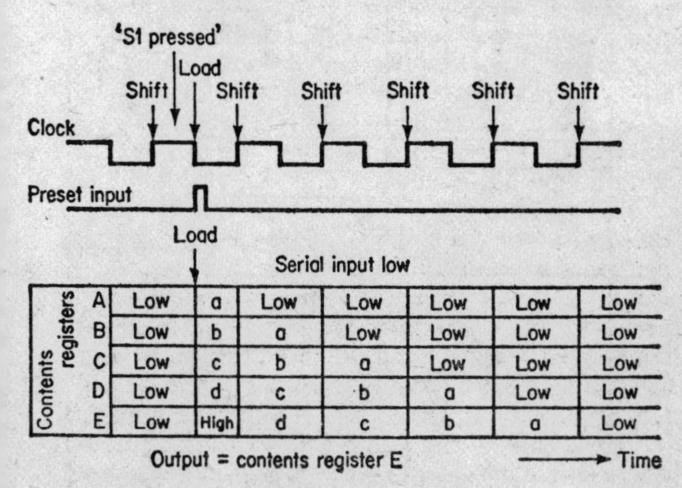

	Low	a	Low	Low	Low	Low	Low
A	Low	a	Low	Low	Low	Low	Low
B	Low	b	a	Low	Low	Low	Low
C	Low	c	b	a	Low	Low	Low
D	Low	d	c	b	a	Low	Low
E	Low	High	d	c	b	a	Low

Fig. 8.3 Shifting sequence in the coder.

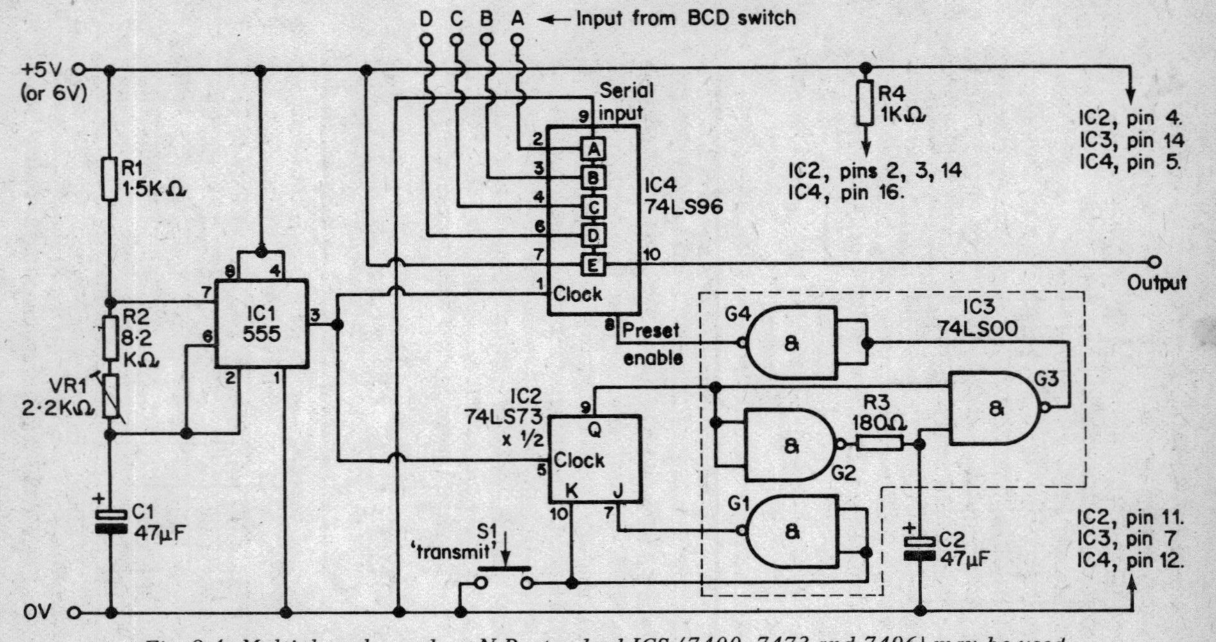

Fig. 8.4 Multiple-pulse coder. N.B. standard ICS (7400, 7473 and 7496) may be used.

or lows in any combination, depending on the code selected, and we represent these by a, b, c, d. These data do not affect the registers yet. Next the operator presses the 'Transmit' button and at the next *low*-going clock pulse the data are loaded directly into the registers (parallel loading). They remain there until the next high-going clock pulse, when they are shifted one step along the chain. At each succeeding high-going pulse they are shifted along until, after 5 shifts, the data has gone and is replaced by all lows. Note that at the first shift register E was 'set' high, because its input is permanently wired to a 'high' voltage. The output of the transmitter is taken from register E. If you read along the bottom line of Figure 8.3, you will find the intended code message: 'start-d-c-b-a', preceded and followed by a continuous 'low' state.

Figure 8.4 shows the circuit details. It operates from a stabilised 5V supply because this is the ideal for the TTL 7400 series i.c.s used. For a portable transmitter you can use a 6V supply (four 1.5V cells) but higher voltages must not be used.

Clock

The 555 timer i.c. is wired as an astable multivibrator. Its output (pin 3) rises, falls as a square-wave output at a rate dependent upon the values of R1, R2, C1 and the setting of VR1. The values give *roughly* equal mark-space ratio.

Input interface

The exact construction of this depends upon the type of control panel preferred. It is simplest to use a switch that gives binary output. These are available as slide switches or thumbwheel rotary switches that can be set to any one of 10 positions. If wired as in Figure 8.5, the voltage on each line is held low by the pull-down resistor, except when connected through the switch to the positive supply. As the switch is rotated or slid from position 0 to position 9, the lines carry voltages equivalent to binary numbers from 0000 to 1001. This allows you to work with up to 10 different commands. If you need more, or if you prefer to use press-buttons, toggle-switches or keys, construct a diode matrix as

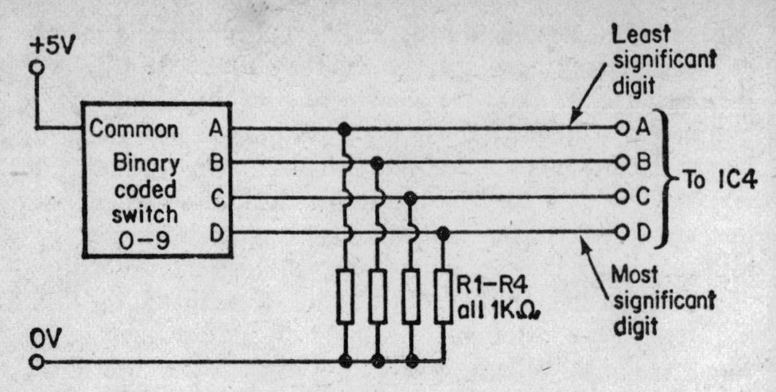

Fig. 8.5 Connecting a binary coded switch to the coder of Fig. 8.4.

in Figure 8.6 to produce the outputs you require. You may need all 16 but, if you do not, you may find that it simplifies decoding problems later if you choose carefully which codes are to correspond with a given set of commands. For example, if all engine-speed codes have digit D high, this makes it easier at a later stage to pick out engine-speed commands from commands of other types.

If you use a binary coded switch, this retains its output until you change the setting of the switch at the next command. If you are using press-buttons or keys, these must be held depressed for long enough *after* the 'Transmit' button has been pressed to allow time for the command to be registered. With the given clock speed, this period is a minimum of about 0.6 second.

Shift register

For this we use a 74LS96 i.c. The low power Schottky TTL is preferred because its low power requirements make it more suitable for battery operation. If you are planning to power your equipment from a mains power-pack, there is no reason why the slightly cheaper standard TTL i.c., the 7496, should not be used. The same applies to the other TTL i.c.s of this circuit. The serial input is wired to common permanently, to reset register A at each shift. The input of register E is wired to V_{cc} (5V or 6V) so that this is set every time the preset

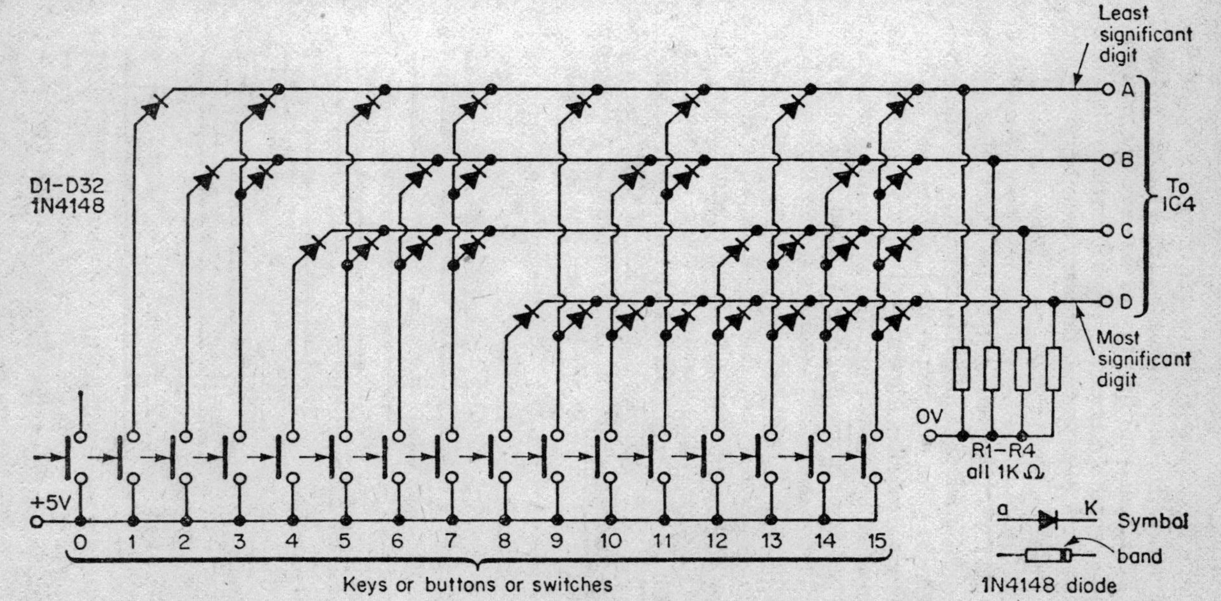

Fig. 8.6 Connecting a keyboard to the coder of Fig. 8.4.

enable is pulsed, that is, when a train of pulses is about to be transmitted. The outputs from registers A to D are not used. The clear input (pin 16) is not used, but must be connected to V_{cc} through a 1kΩ resistor, which it can share with unused inputs of other i.c.s that must similarly be held high.

Preset control

The purpose of this is to generate a brief high pulse on the first low-going clock pulse following the pressing of 'Transmit' button, S1. This must happen once and once only, and no further pulse is allowable during the process of shifting the data out of the register. Figure 8.7 explains the working of this part of the circuit. When S1 is pressed, the K input to the flip-flop (half of IC2) goes low and the input to J (via inverting gate G1) goes high. With the inputs to the flip flop in this condition the output of the flip-flop (Q) goes high at the next *low*-going clock pulse. It will remain high until the next low-going clock pulse after the button is released. As Q goes high, so does one input to G3. The input to G2 goes high and its output falls but, because of R3 and C2, the other input to G3 does not fall quite as quickly. For a brief period both inputs to G3 are effectively high, and its output goes low.

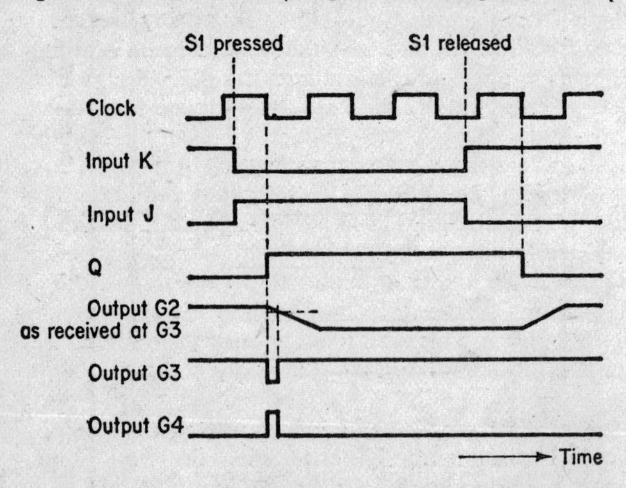

Fig. 8.7 Voltage levels in the coder circuit.

The effect is quickly over and, as the delayed fall to low input takes effect, G3 has one input high and the other low, so its output becomes high again. Thus a brief low pulse is obtained from G3; this is inverted by G4, giving a brief high pulse that enables the transfer of data from the input interface to the registers. After this stage, S1 may be released; Q goes low and, though the output from G2 to G3 rises slowly, the fact that the other input to the gate (fed by Q) is already low means that no low output is obtained from G3. Once the preset enable pulse has been produced, the data is shifted to the transmitter under the control of clock, and it makes no difference if the 'Transmit' button is released. The preset control circuit requires 4 NAND gates, provided by a single 74LS00 (or 7400) i.c.

Transmitter

The output from the coder must next be converted into a form suitable for transmission. Methods of doing this for infra-red radiation, visible light and radio are given in Chapters 6, 7 and 15 respectively. The ultra-sonic transmitter of Chapter 3 can be connected directly to this coder. Simply join the coder output to pin 1 of the i.c. There is no need to remove R1 of the transmitter circuit. Both transmitter and coder should be powered from the same source – a 5-volt regulated mains supply, or (preferred for portability) a 6V battery. There will be a slight reduction in range when the transmitter is powered by 6V instead of 9V, but this has been found to be negligible. On no account must the coder be operated from the 9V supply as this will over-run the TTL i.c.s. With the coder so connected ultra-sound is produced continuously and goes *off* when a high pulse is generated from the coder. In other words, the signal transmitted is the *inverse* of the signal produced by the coder. This is of no consequence, provided it is remembered and the decoding circuit in the receiver designed accordingly.

If it is essential for the transmitter to produce the code exactly as it comes from the coder, the coder output must be inverted before it is fed to the transmitter. This is simply done by using a transistor, connected as in Figure 8.8. Here the 15kΩ resistor is an essential part and must not be removed from the circuit. S1 of the transmitter can be retained to allow pulses to be sent manually for other applications.

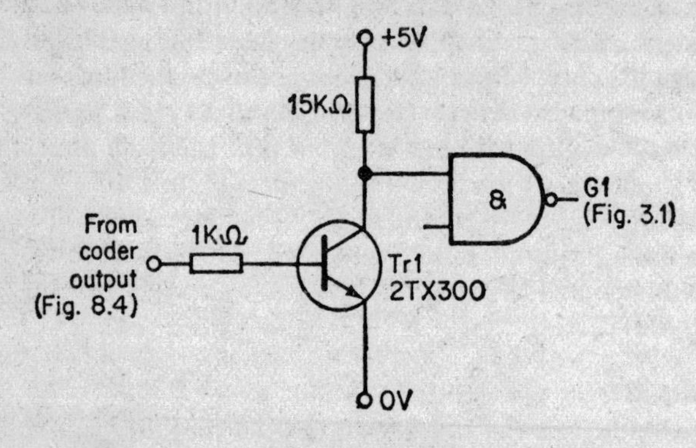

Fig. 8.8. Inverting the output of the coder.

9 MULTIPLE PULSE REGISTER

This circuit (Figure 9.1) is designed to register the values of the individual pulses delivered by the multiple-pulse coder circuit described in Chapter 8. Every command pulse train consists of a half-length 'start' pulse, followed by four full-length code pulses. These may each be high or low depending on the coding.

When the 'start' pulse arrives at the receiver, it causes the trigger pulse generator to produce one very short low-going pulse that triggers the 555 timer (IC2) into action. The relationships between the various pulses of this circuit is illustrated in Figure 9.2. By reading from left to right across the diagram we can follow the various stages of operation. The top line of the diagram represents the command pulse train arriving at the receiver. The 'start' pulse is shown high (S), the code pulses, d, c, b, a, (transmitted in that order) may be low or high, and are followed by a low state that lasts until the next command train arrives. In the second line of the diagram we see the brief trigger pulse which occurs on the leading edge of the 'start' pulse. There are other trigger pulses later (indicated by '?') if a low pulse in the code is followed by a high one, but these have no effect once the 555 has been triggered. The 555 is connected as a monostable multivibrator and its output goes high until half-way through the pulse representing digit a.

The output from the timer is fed to a double-pulse generator that produces a high 'count' pulse as the timer output goes high and a high 'stop' pulse as the timer output goes low (Figure 9.2). The 'count' and 'stop' pulses are on separate lines. The purpose of the 'count' pulse is to synchronise the clocking of the shift register in this circuit with the clocking of the shift register of the transmitter circuit. The clocking pulse comes from IC4, a divide-by-ten i.c. that is driven continuously by the clock. The clock is made from two NAND gates and has a frequency ten times faster than the clock of the transmitter (Figure 8.5). The D output of

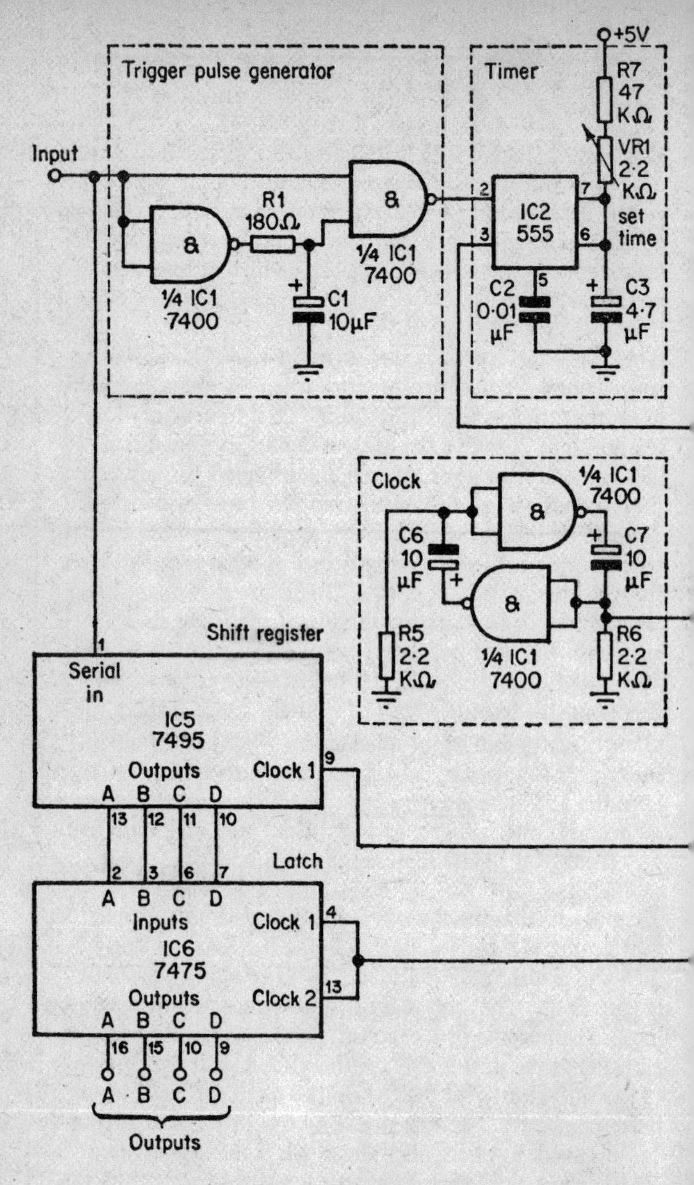

Fig. 9.1 Multiple-pulse register.

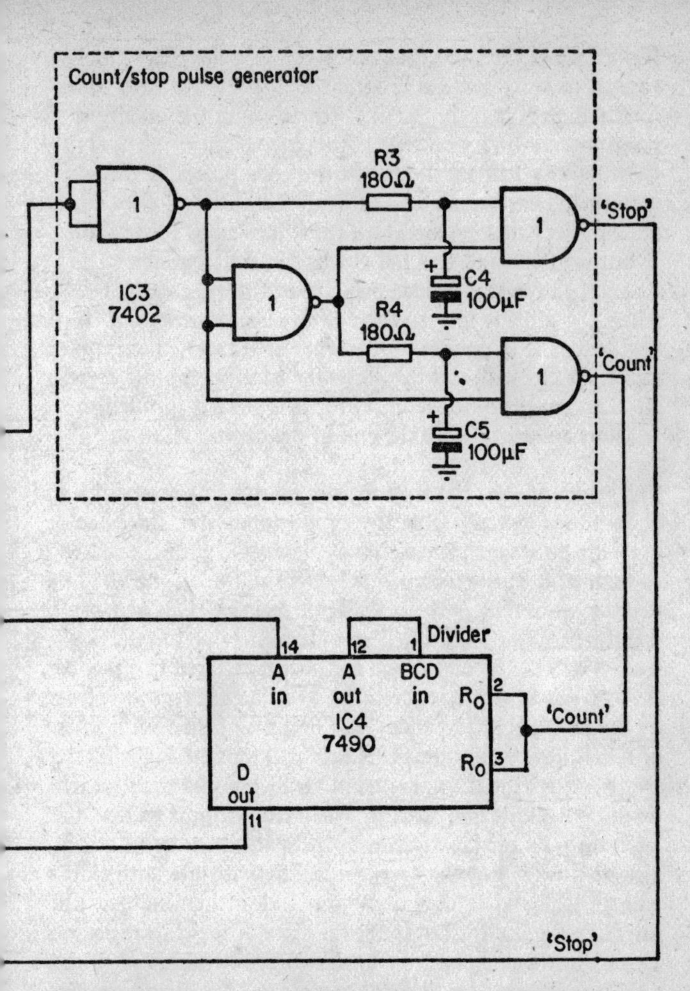

To V_{CC} (+5V):
IC1, pin 14
IC2, pins 4, 8
IC3, pin 14
IC4, pin 5
IC5, pin 14
IC6, pin 5

To ground (0V):
IC1, pin 7
IC2, pin 1
IC3, pin 7
IC4, pins 6, 7, 10
IC5, pins 6, 7
IC6, pin 12

To V_{CC} through 1KΩ resistor:
IC5, pins 2, 3, 4, 5 (parallel inputs)

IC4 thus has the same frequency as the transmitter clock, though its duty-cycle is different, being low for four times longer than it is high. At the 'count' pulse the divider is reset and continues counting from zero. Since the 'start' pulse is only half the length of the code pulses, it reaches zero again exactly half-way through the period of each of the code pulses. As the count changes from 9 back to zero, the D output falls, and this fall clocks the shift register. The diagram shows the divider being reset from a count of 6, when D is already low, so there is no pulse on resetting. If the divider is at counts 8 or 9 when it is reset, the shift register is clocked, and this shifts the 'start' pulse into register A. This makes no difference, for it is shifted through the registers and lost before the end of the operation.

The serial input of the shift register receives the start pulse and code pulses directly from the circuit input and, since the clocking pulses are synchronised to coincide with the centre of each code pulse, the code pulses are shifted as shown in the diagram until they each are held in a register. This occurs after 4 shifts (or after 5 if there is an initial shift due to resetting from 8 or 9). The code must next be transferred to a second set of registers where it can be held (or latched) until replaced by the next command code. The outputs of the four latches of IC6 follow their inputs if the clock inputs are high. There are two clock inputs, operated in parallel in this circuit, each input controlling two latches. When clock inputs go low, the latch outputs remain latched in the state they are in at the instant clock goes low. They remain in that state until clocks go high again, when they once more follow their inputs. The brief 'stop' pulse thus captures the state of the shift registers. This loads the code into the latch registers, where it remains until it is replaced by the next command. The time from the pressing of the 'transmit' button until the appearance of the code at the latch register outputs is approximately 1 second. This is fast enough for most applications. If faster response is essential the clock rates may be increased, though this could lead to difficulties in adjusting the system and might make it more susceptible to mis-timing should the values of electrolytic capacitors change, as they often do with age, especially

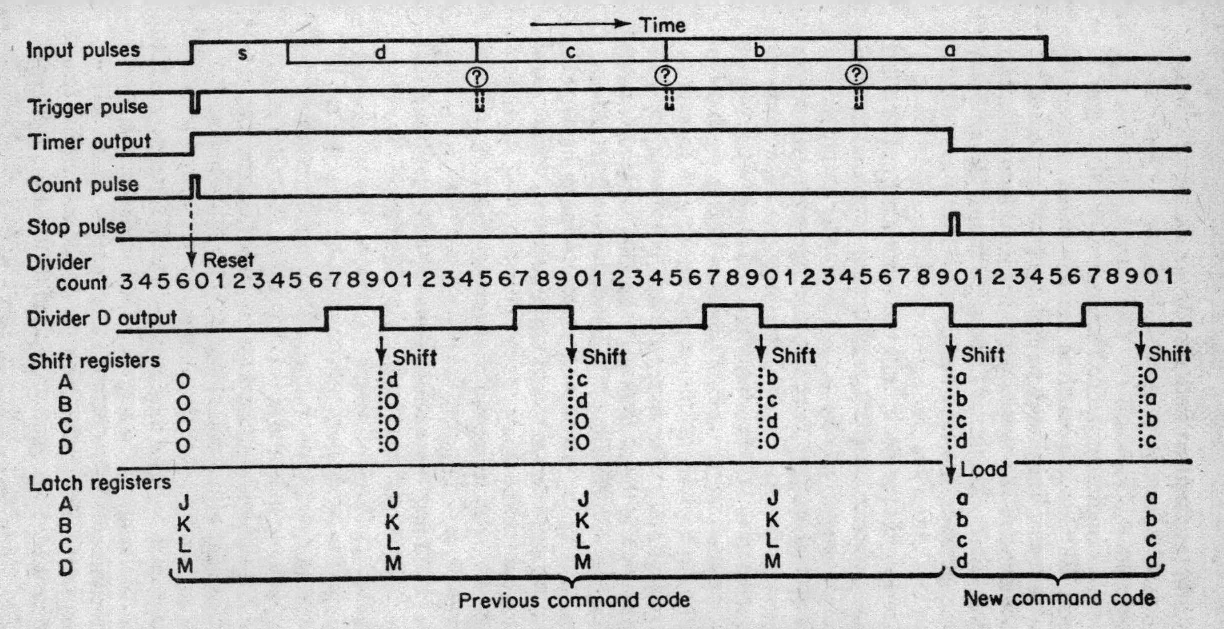

Fig. 9.2 *Waveforms of the multiple-pulse register circuit.*

when left unused for a period.

Construction

The unit requires only 6 i.c.s together with a few other components, so can easily be accommodated on a small circuit-board. The prototype used a piece of strip-board 50 holes long and 18 strips wide, the i.c.s being arranged in two rows of 3. The circuit can be powered from a 5V regulated power-pack, or from a 6V battery.

First assemble the clock, using two of the gates of IC1. Next connect the divider (IC4), remembering to ground the two 'Reset 9' inputs (pins 6 and 7). Temporarily ground the two 'Reset 0' inputs (pins 2 and 3) so that the clock and divider can be tested. Outputs are most conveniently tested by connecting them to an LED in series with a 180 ohm resistor which is itself connected to ground. Connect an LED to the D output of the divider; it should flash regularly, about 15 times in 10 seconds (1.5Hz). If another LED is connected to the output of the clock in the transmitter, this should be seen to flash at the same rate. The rate of the transmitter clock should now be adjusted, using VR1 *of the transmitter*, until the two LEDs flash at equal rates, as near as can be judged by watching them. Since the transmission cycle occupies only 4 clock pulses, slight discrepancies are of no importance.

Next build the trigger pulse generator, using the two gates remaining unused in IC1. Wire up IC2 and the connection between it and the trigger pulse generator. The trigger pulse is too short to be observed using an LED, but its action can be checked by testing the output of the timer. Immediately a high pulse is applied to the circuit input, the timer output should go high and remain high for approximately 1 second. Adjustment of the timer pulse length will be done later.

The count/stop pulse generator uses all 4 NOR gates of a 7402 i.c. The values of C4 and C5 are not critical; any surplus capacitors with values between 22μF and 150μF can be used. Connect 2 LEDs to this generator to check its action (see Figure 9.2). Next connect the 'count' output to the 'Reset 0' input of IC4. Now an LED connected to the D output should flash regularly, except when the arrival of a high pulse at circuit input causes the timer to operate and to generate the 'count' pulse that resets IC4. This shows as an interruption in the regularity of the flashing rate. It should resume flashing regularly about 0.7 seconds later. If an LED is connected to the timer output, we are ready to adjust the length of the timer pulse. Simply apply a high pulse to the circuit input; timer output should go high immediately and stay high until *just after the* fourth flash from the LED connected to divider D output. Adjust VR1 (*receiver*) until the timer LED goes out just after the fourth flash of the divider LED.

Complete the circuit by connecting the shift register and latch i.c.s, and it will then be ready for final testing. For this, remove all test LEDs from the circuit, and then connect four LEDs to the latch outputs. Connect the output of the transmitter by wire to the input of the receiver. Power both transmitter and receiver from the same source for the purposes of testing. When the circuit is switched on, one or more of the LEDs may light, at random. Select control position '4' (= binary 0100; so c = 1, a = b = d = 0). Press the 'transmit' button. About one second later the LED connected to register C should light, and the other three LEDs should be dark. If more than one LED is lit, something is wrong with the timing (you may be catching the 'start' pulse), or with the wiring associated with the control switch, the shift registers or the latch. If one LED lights but it is the one connected to latch B, this indicates that latching has occurred *before* the final fourth shift. Lengthen the timer pulse by adjusting VR1 of the receiver. Conversely, if the D LED lights, reduce the length of the timer pulse. At each adjustment, press 'transmit' to repeat the test. Next select control position '9' (= binary 1001; so a = d = 1, b = c = 0) and repeat the test. If LEDs A and D light this is a sure indication that the pulse train has reached exactly the correct registers at the time of latching. Finally run through all control positions to check for correct action. The receiver is now ready to be connected to a decoder designed to interpret the command codes and relay corresponding instructions to the controlled device. Details of how this is done are given in Chapters 6 and 11.

10 CHANNELS AND DECODING

When designing a remote control system, three points must be carefully considered:

(1) How many devices are to be *independently* controlled?
(2) How many devices are to be *simultaneously* controlled?
(3) Which devices require digital control and which require analogue control?

Independent control means that the device may be turned on or off, or its speed altered without affecting any of the other devices in any way. Sequential control (Chapter 5) whether achieved electronically or by means of ratchet mechanisms is not independent control, because *all* devices come into action in sequence, even though the unwanted functions may be passed over so quickly that they have no time in which to take much effect. We must consider the sequential control circuit itself as the controlled device, so that we have a *single* device and the question of independence does not really arise. It is a simple system with regard to transmission and operates through a single channel.

Simultaneous control means that while we are controlling, say, the position of a rudder, we may also vary the speed of the propulsion motor. Each device may be controlled at any instant, and does not have to wait its turn. In such a system we need a separate channel for each device.

Digital control means that a device may be switched on or switched off but has no in-between state. With analogue control we are able, for example, to vary the speed of a motor smoothly (or at least set it to one of a graded range of speeds) and not simply turn it on or off. The matter of analogue control is dealt with in Chapter 11. Each device that is simultaneously controlled, either in a digital or an analogue way, requires one channel.

To make the idea of channels clearer, refer to Figure 10.1.

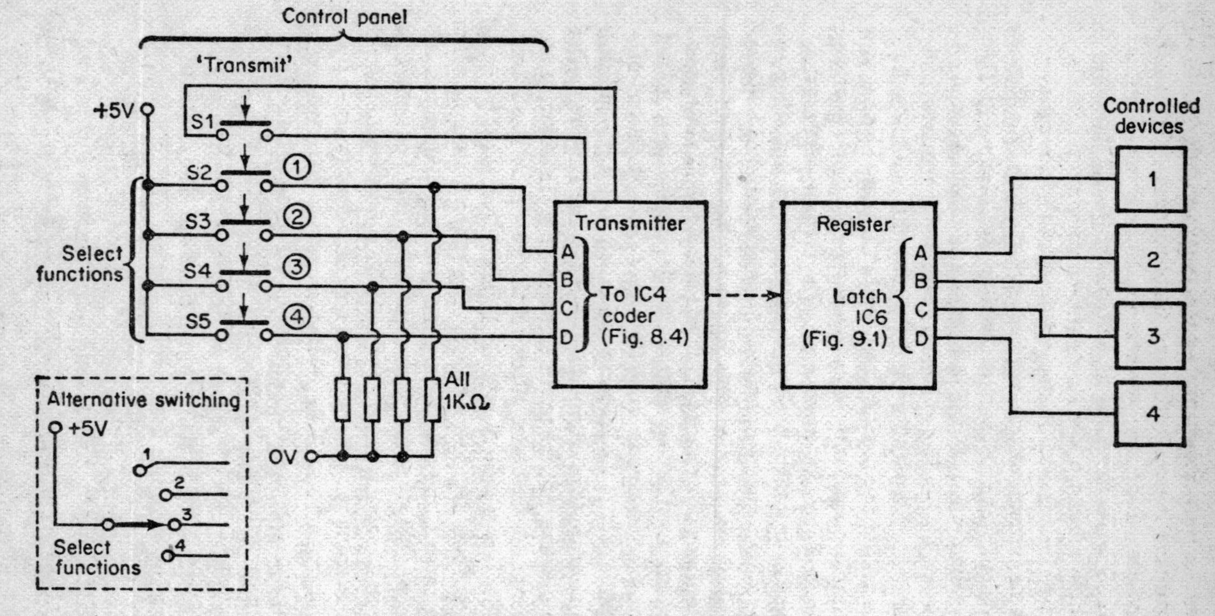

Fig. 10.1 Four-channel system.

Here we show a 4-channel system, allowing 4 devices to be controlled independently and simultaneously. To switch any given device on, we press the corresponding buttons (S2–S4) and the transmit button. For each function button pressed, a 'high' output eventually appears at the corresponding latch in the register circuit of the receiver. These outputs can be used to switch on the devices required, using the circuits described in Chapter 11. To switch a device off, press the buttons of functions required but not the buttons of functions no longer required, and transmit. This kind of circuit, using the multiple pulse transmitter and register described in Chapters 8 and 9, is versatile enough for many purposes and there is no complex decoding to be done. The control panel of the transmitter can be set out in various ways, according to the kind of model or equipment being controlled. Figure 10.1 shows separate keys for each function, with the alternative of a rotary switch, which is simpler to instal and for any given setting turns *one* device on and turns *all* the non-selected devices off. Obviously the applications of remote control are so many and varied that it is impossible to do more than outline the principles that may be applied to the design of the reader's own system. However, as an example of the degree of control that can be obtained, even with a simple 4-channel system, Figure 10.2 shows how a model motor-boat may be controlled, using relays to turn the motor on and off, and for reversing motor current, and a pair of electromagnets to turn the rudder either to left or right of its neutral central position. Normally the rudder would be held in a central position by hydrodynamic forces (assisted by light rubber bands if the bearings are stiff). The control panel has two rotary switches, one for motor-speed and direction, one for rudder, and a transmit button to send each new command. Each device or function is independently and simultaneously controlled.

In Figure 10.3 we see a one-channel system that provides analogue control of the speed of a motor. It could equally well be used to control some other function, such as the temperature setting of a thermostat. In this system the 4-digit

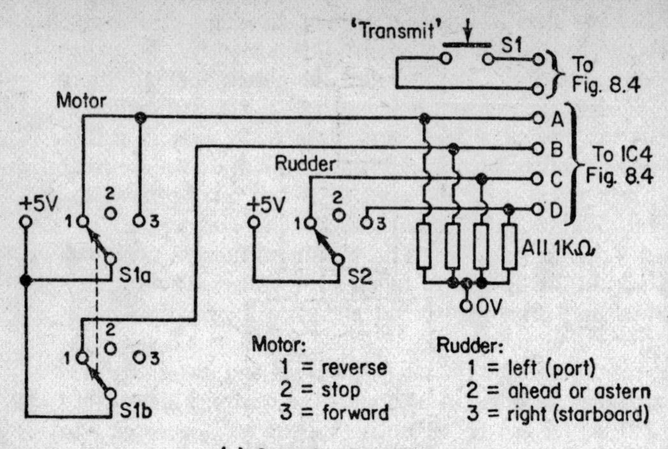

(a) Control panel switches.

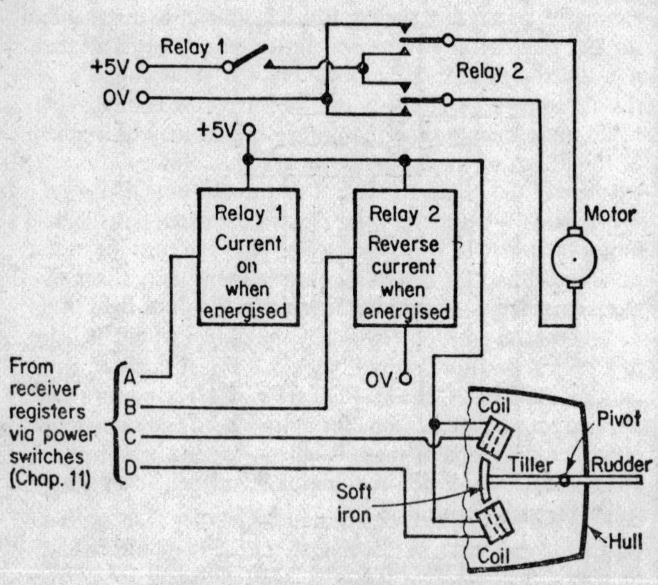

(b) Connections to controlled devices.

Fig. 10.2 A four-channel system applied to the control of a model boat.

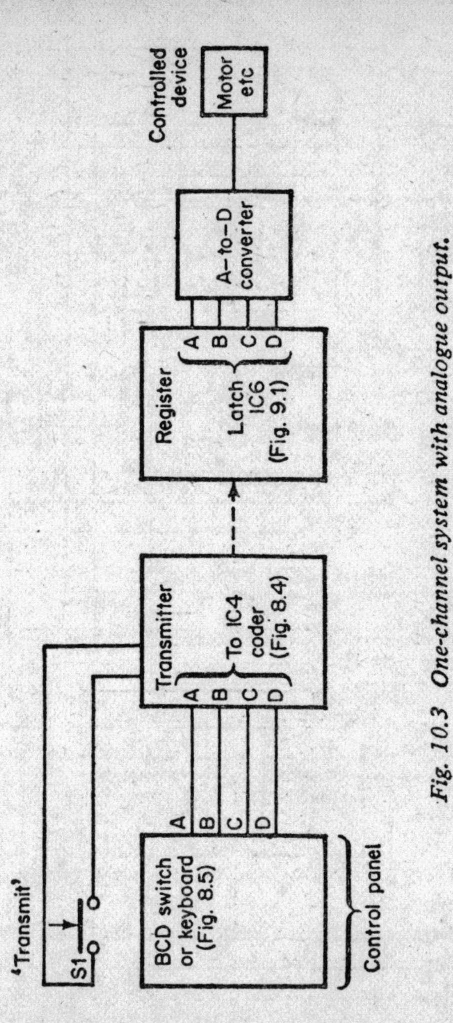

Fig. 10.3 One-channel system with analogue output.

code is thought of as a binary number, rather than as four separate commands. According to the value of the binary number transmitted, the motor is made to turn at one of 16 different speeds. An essential part of this system is the digital-to-analogue converter. This accepts the binary number as input, but its output is a voltage or current, the value of

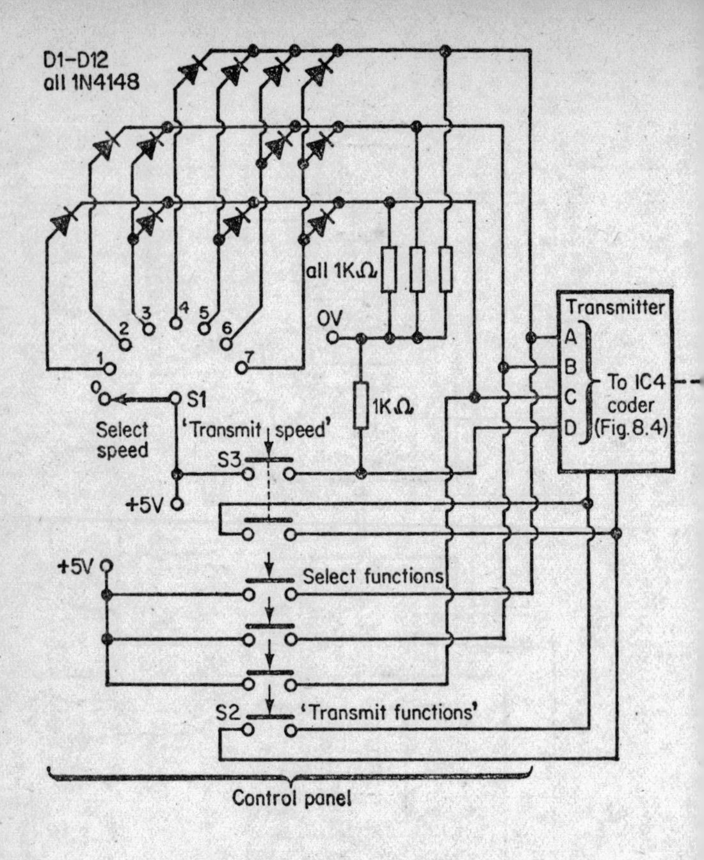

Fig. 10.4 Four-channel system with analogue output.

which is determined by the value of the digital input. The construction of digital-to-analogue converters is described in Chapter 11.

With the system shown in Figure 10.3, only one device can be controlled, though with a little more complexity of decoding it is possible to provide analogue control (Chapter 11) of one channel, and digital control of three other channels, all with independent and simultaneous control. Only 4 digits are required, so this system can make use of the multiple-pulse

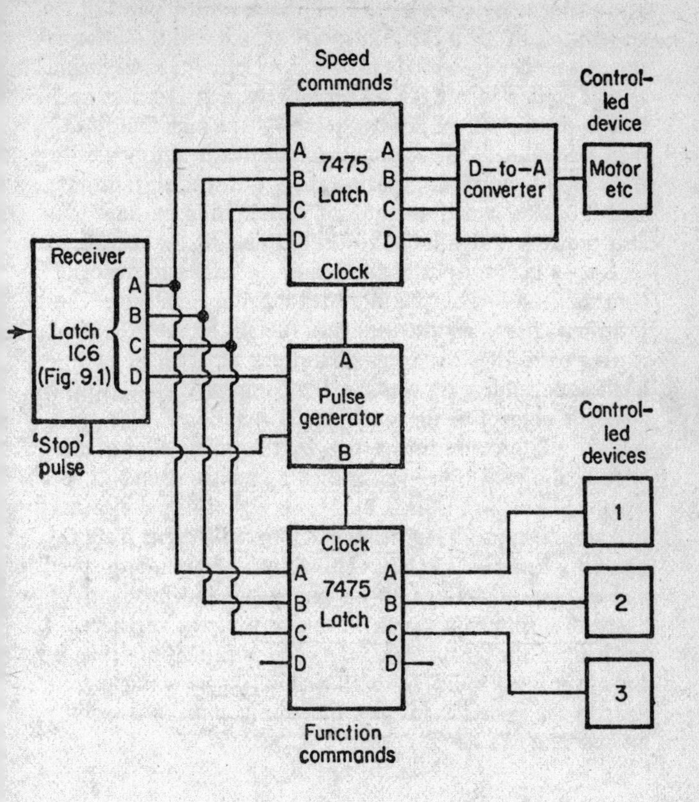

transmitter and register of Chapters 8 and 9, yet the more efficient decoding system allows for the wider scope of control. It is given as an example of one of the many ways in which a 4-digit transmission can be used.

In the system shown in Figure 10.4, binary numbers 1000 to 1111 (8 to 15 in decimal) represent commands to control motor speed, or some other analogue function. Binary numbers 0000 to 0111 (0 to 7) represent digital control of 3 other channels, one channel being controlled by each digit.

The analogue function is selected on the control panel of the transmitter by an 8-position rotary switch and the information is encoded by a diode matrix in exactly the same manner as that used in Figure 8.6. Alternatively, a BCD switch could be used in any one of positions 0 to 7. The high first (D) digit common to all analogue commands is automatically transmitted each time, because there is a special 'transmit speed' button which, in addition to initiating transmission also provides a high input to the D line. At the receiver register, a high output at the D latch indicates an analogue command. A low output indicates a digital command. The D output goes to a pulse generator (details Figure 10.5) that is used to register the command in one or other of two latches, depending on whether the command relates to the analogue control of the three digital functions. This pulse generator is brought into action by the falling edge of the 'stop' pulse, which has just caused a new command to be registered in IC6 (Figure 9.1). If D is high, this pulse causes the speed command register to latch to the 3 digits A, B, C, currently being held in IC6. This system allows motor speed and 3 other functions to be controlled independently. Control is simultaneous in the sense that the registers 'remember' the most recent command until they are changed, though it is not possible to alter an analogue command at *exactly* the same instant as a function command is being altered, and *vice versa.*

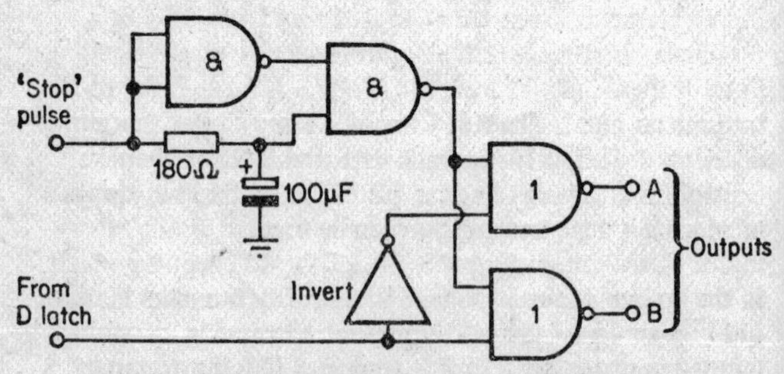

Fig. 10.5 Details of the pulse generator of Fig. 10.4. The INVERT gate can be a spare NAND or NOR gate with its inputs wired together.

If we are prepared to sacrifice simultaneous control even further, we can use the same 4-pulse system to control even more functions. The 3 digits used for function control correspond to eight different binary numbers and can be used to control up to 8 different functions if decoded. A single i.c., the 7442 binary to decimal decoder, is available to do this (Figure 10.6) it can accept a 4-digit binary input, though if used in conjunction with the circuit of Figure 10.4, only three digits would be used, input D being grounded. Normally any given output is high but, when a particular number is input in binary, the corresponding decimal output goes low. If one of the 'disallowed' binary numbers, say, 1101 (= 13) is input, the decimal outputs are all high. The outputs can be used to switch on or off any number of devices up to 10. We have to decide whether we wish the device to be switched on for only as long as the corresponding binary number is in the latch, or whether we want to have toggle-action the device remaining switched on until a counter-command is given to switch it off again. Before considering how this can be done, the other decoders illustrated in Figure 10.6 should be mentioned. The 4028 is a CMOS equivalent of the 7442, performing a similar function, except that its decimal outputs are normally *low* and go *high* when the corresponding binary number is input. An advantage of the 4028, common to all CMOS i.c.s, is the low current consumption and the ability to operate at any voltage in the range +3V to +15V. The 74154 is similar to the 7442 except that it can decode all binary numbers from 0000 to 1111 and has 16 output terminals. It also has strobe input terminals; when decoding, both of these must be low, and if either is made high, all 16 outputs go high. The 4514 and 4515 are CMOS 16-output decoders. The 4515 operates like the 74154 when its 'follow' input is high and its 'inhibit' input is low: outputs are normally high and go low when addressed by the binary input. If the 'inhibit' input is made low, *all* outputs go high. If the 'follow' input is made low, the i.c. acts as a latch; its outputs remain in the state they were in just as the follow input was made low. This feature can be used to latch commands until new commands arrive. When a new command arrives a short high pulse causes the new commands

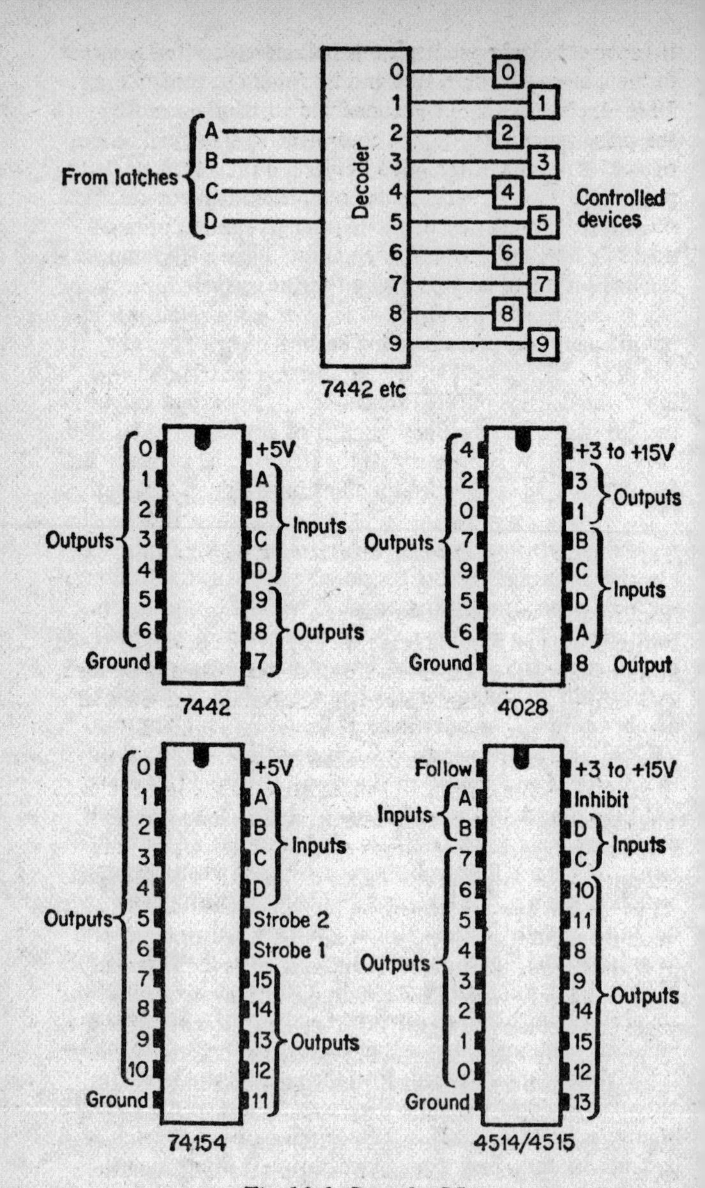

Fig. 10.6 Decoder ICs.

to replace the old one. Thus this i.c. can replace the function of the latches of Figure 10.4 and the 7442 decoder of Figure 10.6. Its 'follow' input is connected to the B output from the pulse generator, for it is equivalent to the 'clock' input of 7445. The use of the 4515 in place of the 7475 and 7442 saves current and reduces the amount of wiring required. The 4514 is very similar to the 4515, except that outputs are normally *low* and go *high* when addressed; a high input to the 'inhibit' terminal causes all outputs to go low.

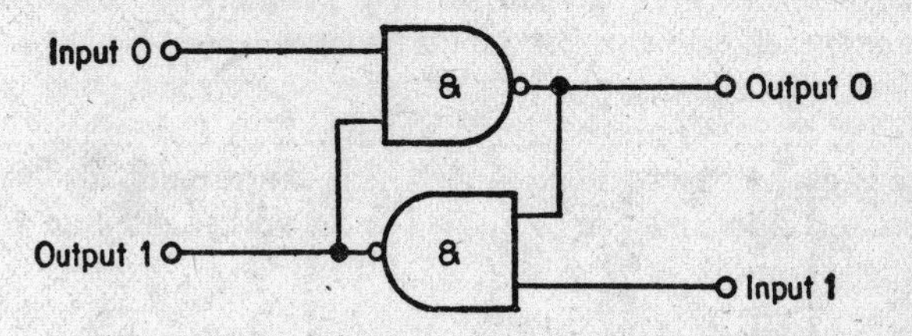

Fig. 10.7 A bistable for obtaining toggle action.

The simplest way to obtain toggle-action is to operate the device through a bistable (Figure 10.7), which is usually made from two NAND gates. Both inputs should normally be high. One output is high, the other low. The bistable can exist in either one of two stable states. If a low pulse is applied to input 0, output 0 changes from low to high (unless it is already high) and output 1 changes from high to low (unless already low). Further low pulses on input 0 have no effect, but a low pulse on input 1 causes output 1 to go high, and output 0 to go low. Further pulses on input 1 have no effect. Thus the bistable can 'remember' on which input line it most recently received a low pulse. If input 0 is fed from output 0 of the decoder (Figure 10.6) and input 1 is fed from output 1, a command signal '0000' causes the bistable to assume one state and a command signal '0001' causes it to change to the opposite state. By setting and resetting the bistable we may arrange to turn a motor on at command 0000 and turn it off at command 0001. Similarly the other outputs of the 7442 may be paired to provide toggle action on a total of 5 devices, if required. Since the outputs of a bistable are of opposite sense, a single bistable can turn one device off and switch another on simultaneously.

Wiring up bistables is tedious but fortunately i.c.s are made that contain several bistables ready-built. The 74118 (Figure 10.8) has six bistables. Each has a separate 'set' input, but they all share a common 'reset'. To use this we could have six different command functions to turn devices on. A single command (say 0000) would turn all off together. This is a useful feature in some applications. The 74279 has four bistables, each with separate 'set' and 'reset' inputs. Bistables 1 and 3 each have *two* 'set' inputs (using a 3-input NAND); the effect of this is that the bistable is set when a low pulse is applied to *either one* of the set inputs. This feature could prove useful on occasions though, if it is not required, the two 'set' inputs may be connected to each other and used as a single input.

So far, we have seen how the control systems are able to operate latches and bistables under remote control. In the next chapter we examine the final connecting link, the power switch, by means of which a remote device is eventually brought under the control of the operator.

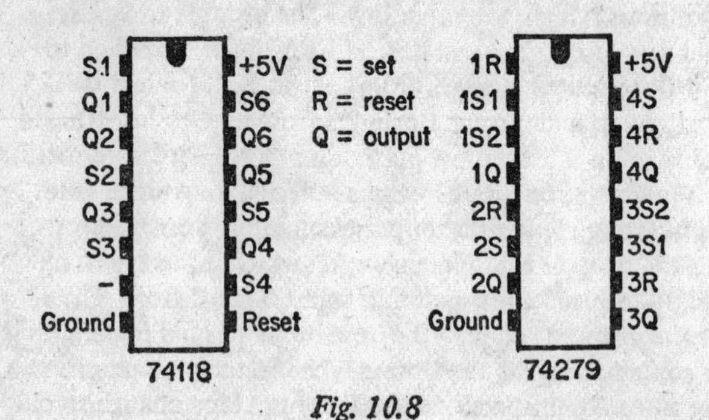

Fig. 10.8

11 POWER SWITCHING

The output of most of the remote-control receivers described in this book comes either from a TTL or CMOS gate or from the collector terminal of a low-power transistor. This output may be used directly, or fed to a decoder which, likewise has only a low-power output. The output may be used to light an LED or perhaps a small filament lamp, but can not do much more. The maximum currents that may be taken from such outputs are listed below:

Source	*Maximum current* (mA)	*Normal operating voltage* (V)
Standard TTL gate (74 series)	16	5
Standard TTL buffer gate (e.g. 7440)	50	5
Low power TTL gate (74L series)	0.18	5
Low power Schottky TTL gate (74LS series)	0.36	5
CMOS (BE series)	4	5
	11	10
Low-power transistor (ZTX300)	500	up to 25

The current available at the collector terminal of a transistor depends on how much current the transistor can safely conduct when switched on. In Figure 11.1a, all current passing through the load also passes through the transistor. The maximum current allowed is a few hundred milliampere if the transistor is of the low-power type. If the load requires only a low current it may be necessary to connect a resistor in series with the load to limit the current flowing. If a current greater than a few hundred milliamperes is required, we simply substitute a transistor of higher power rating. Examples are the BD131, which can handle up to 3A, and the 2N3055 which takes up to 15A. The circuit in Figure 11.1a powers the load when the transistor is switched on (high

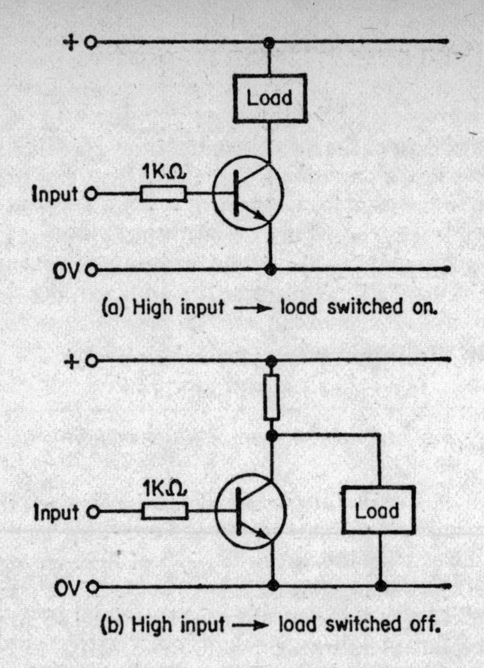

Fig. 11.1 Using a transistor as a switch.

input). The circuit of Figure 11.1b powers the load when the transistor is switched off (low input). When the transistor is on, all or most of the current flows through the transistor, by-passing the load. There is still a current flowing through the load, though normally this is small, for the effective resistance of the switched-on transistor is much less than the resistance of the load. If the load is a lamp, the current may be insufficient to make the filament glow. With other kinds of load even a small current may be unacceptable; the solution is to use the switching circuit of Figure 11.1a, but with an inverted input. If a spare logic gate is available this may be used to invert the load. If not, a transistor can be used as inverter as in Figure 11.2.

TTL i.c.s can not be operated above 6V, but if a load requires higher voltage, it can be fed from a separate supply, as in

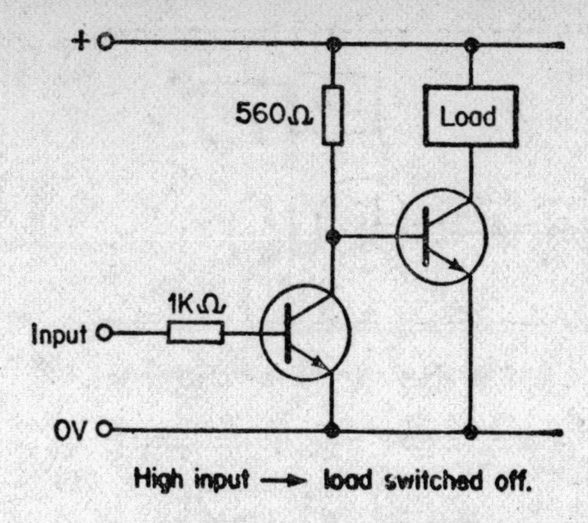

Fig. 11.2 Another transistor switch.

Figure 11.3. Both sides of the circuit share a common 0V rail. The voltage on the load side must not exceed the maximum collector-emitter rating (V_{CEO}) of the transistor. When using high voltages, high currents, or both, the maximum power rating of the transistor must also be taken into account. Typical values are 300mW for the ZTX300, 15W for the BD131 and 90W for the 2N3055. When operated at high power the transistors must normally be mounted on an adequate heat-sink. The only exception is if the transistor is switched on only briefly and with long 'off' periods between. Then the small quantity of heat generated during the 'on' period has time to escape during the 'off' period, without need for a heat-sink.

A precaution must be taken when switching *inductive* loads, such as motors, bells, buzzers and relays. In these the current passes through one or more electromagnetic coils. When current is switched off, the sudden collapse of the magnetic field causes a high e.m.f. to be induced in the coils. This e.m.f. may be tens of volts in magnitude, even though the supply current to the load was well below 10V. The e.m.f. causes a high current to flow through the coils and, of course, through transistors or i.c.s that are connected to the coils. Since the induced current is always in a direction *opposite* to that of the supply current it can do a great deal of damage

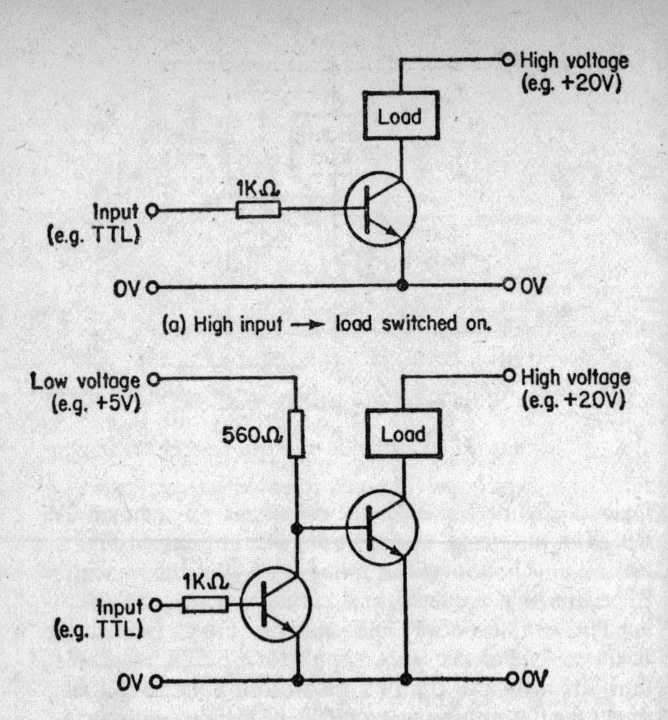

Fig. 11.3 Switching a load that operates from a higher voltage.

in an instant. To guard against this we connect a diode across the terminals of any inductive load (Figure 11.4). This discharges the reverse current safely.

Mention of relays reminds us that here is yet another means of switching power to heavy loads. Relays are relatively expensive, bulky and slow-acting compared with transistors but, for heavy loads or when the load is to be driven by alternating current, a relay is to be preferred. Figure 11.5 shows the complete circuit of an ultra-sonically controlled relay. This could be used for controlling room lighting or for switching on any other mains-powered device, such as a radio set, an electric fan, or a film projector. The use of a flip-flop (4027) gives it toggle action: press 'transmit' to

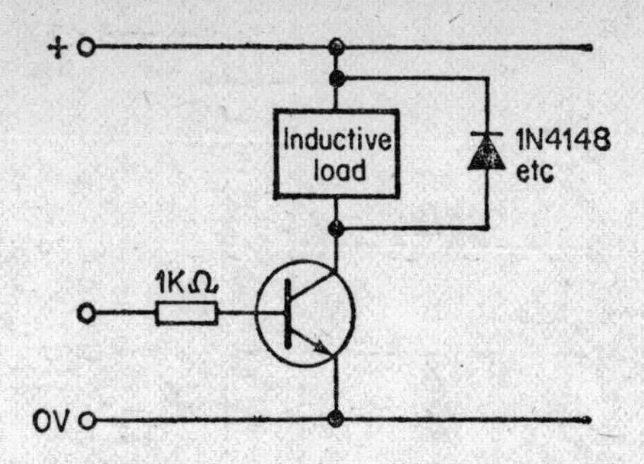

Fig. 11.4 Protective diode in use with an inductive load.

switch on; press again to switch off. By suitable choice of relay, this circuit could be used to switch on one device while simultaneously switching off another. Wired between the pick-up cartridge and amplifier of a hi-fi system, the circuit could be used for instantly muting the system when, for example, a telephone call has to be made. In this application a heavy-duty relay is not required. Instead one could use one of the small relays in 14-pin d.i.l. cases (the same size and shape as a TTL or CMOS i.c.). These can be used to switch voltages up to 100V and currents up to 0.5A. Their small size makes them particularly convenient for model-control circuits and it is convenient to be able to mount them on the same circuit-board as the i.c.s. Another advantage is that they can be powered directly by the output of a standard TTL gate. One brand of d.i.l. relay has the protecting diode already built-in, but it is wise to check for this feature before using relays of other makes.

Controlling motor speed

So far we have been concerned only with functions that are to be either on or off. This is not sufficient for the control of models such as boats and vehicles for which control of speed is almost an essential item. Speed is a quantity that can be varied smoothly from zero up to some maximum value; it can have all possible values between these two extremes. We call this an *analogue* quantity. Its value may be represented in analogue form as, for example, when the wiper of a potentio-

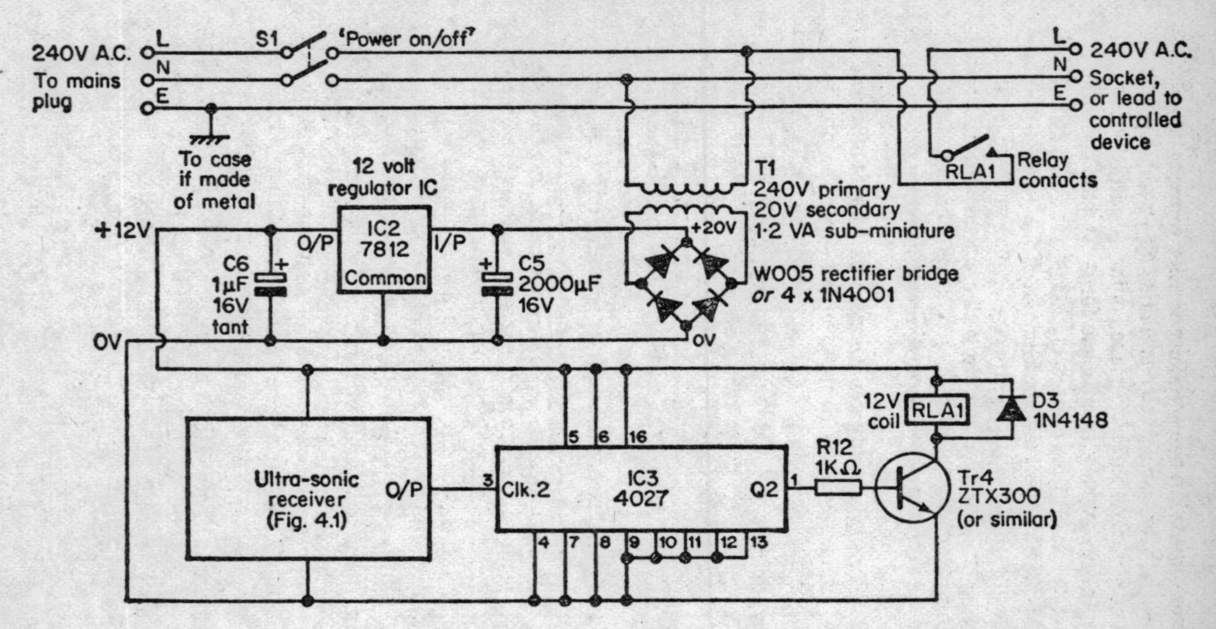

Fig. 11.5 Ultra-sonically controlled relay for switching mains-powered devices.

meter is turned to one particular position of the many possible ones. Its value can also be *represented by* a number – very often a binary number to make coding simpler. Such a number is not truly an analogue for it can have only the discrete values of the integers (1, 2, 3, 4, . . . , etc.) and not all the possible values between them but, if the number of integers on the scale is sufficiently large, a numerical value can be considered to be an analogue for all practical purposes. Speed or other analogue quantities can generally be represented as control signals in one of three different ways:

(1) Pulse *length* indicates the analogue value. For example, the longer the pulse the greater the corresponding speed. A series of pulses of differing length can be used to transmit commands that control the speed of the engine and the settings of the rudder and ailerons of a flying model aeroplane. This system is widely used and is known as *digital proportional control*, since it uses a series of pulses (digits), with lengths proportional to the analogue values required. The train of pulses may also include some of standard length for actuating on/off functions such as cutting out the engine completely, or switching on landing-lights. A control system of this kind was featured in *Practical Electronics*, and has been republished as a booklet by Maplin Electronics Ltd. The advantage of the system is that it enables smooth, proportional control of all functions, which is invaluable for applications such as model aeroplanes though less important in many other applications. The disadvantage is that it requires servo-mechanisms for all analogue functions, and to construct these requires skill. Ready-made servo-mechanisms can be purchased but are expensive. Circuits can be simplified by using the NE544 i.c. which is specially designed for decoding digital proportional transmissions.
(2) A *binary number* indicates the analogue value. This is transmitted as a series of pulses coded, for example, as described in Chapter 10. This method is described in detail below.
(3) The required analogue value itself is not transmitted;

instead the receiving circuit is commanded to *increment* or *decrement* by one step the value currently held there. In other words, the two possible commands are 'increase speed' or 'decrease speed' (or 'turn more sharply left' and 'turn less sharply left'). The commands are repeated until the required speed (or amount of turn) is reached. The system to be described in Chapter 13 has three analogue outputs of this kind, each operating over a 32-step range, which gives as fine a degree of control as is required for almost all purposes. A similar action may also be obtained by a simple modification of the system next to be described.

Digital speed control

This was referred to in paragraph (2) above. The multiple-pulse transmitter produces a 4-digit binary number (though it could be extended to 8 or even more digits if required). The code is transmitted and eventually appears in the latch registers of the receiver. The next step is to convert this digital representation of speed into an analogue form. The circuit for doing this, a digital-to-analogue converter, is shown in Figure 11.6. It uses a CMOS i.c. because current from the outputs can be summed simply by feeding them through resistors to the base of TR1. This does not work with TTL. Because we are taking outputs from TTL (the latch of the decoder circuit) and using these as inputs to CMOS we need the pull-up resistors R5–R8 at the interface. The 4050 i.c. contains 6 non-inverting buffers; for a high input, each gives a high output. A high output causes a current to flow through the resistor to the base of TR1, the amount of current being inversely proportional to the resistance. With the values shown in Figure 11.6, the total current is proportional to the 4-digit binary number represented by the latch outputs, D being the least significant digit. Resistors with values 50kΩ and 400kΩ are not available in the standard series, but can be made up by joining two resistors in series. In practice, resistors of values 47kΩ and 390kΩ are normally close enough to the required values. TR1 is connected as an emitter-follower, and so the potential across R9 varies in proportion to the total base current of TR1. The variable

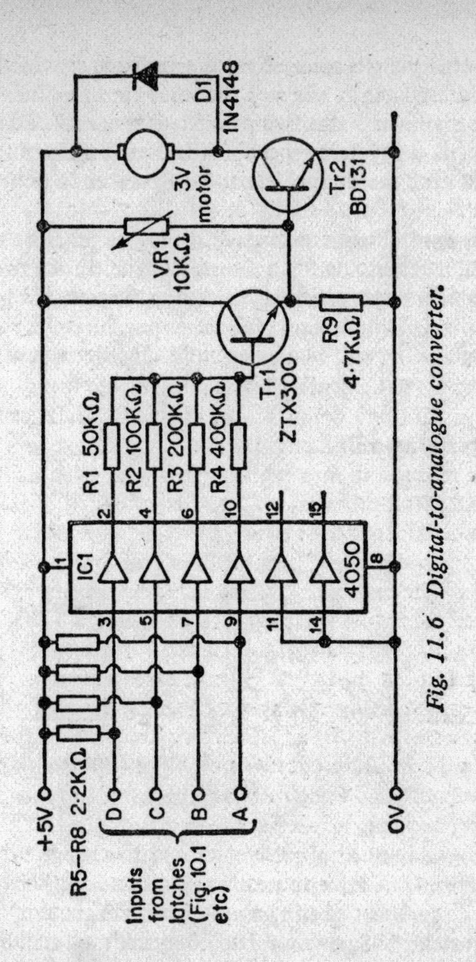

Fig. 11.6 Digital-to-analogue converter.

resistor VR1 supplies sufficient base current to TR2 to keep the transistor switched on, and the motor running at its lowest required speed even when all latches are zero. In any other condition, additional current flows from TR1, turning TR2 more fully on and thus increasing motor speed. The amount of increase is proportional to the binary number stored in the latch. VR1 can be replaced by a fixed resistor if the same minimum speed is always required.

If only 3 binary digits are used for the analogue control, line D is made available for logical purposes (p. 65). The connections to pins 9 and 10 are omitted, as are R4 and R8. It is advisable to try out any proposed circuit on a breadboard to ascertain the most suitable values of resistor to be used.

As well as controlling motor speed it may be required to control motor direction. The simplest method is to use a relay (Figure 11.7), controlled through a separate channel from that controlling speed. It is then possible to control both direction and speed independently. If space is restricted and weight has to be kept as low as possible, use one of the d.i.l. relays with two double-throw contacts. These can be driven directly by a TTL output.

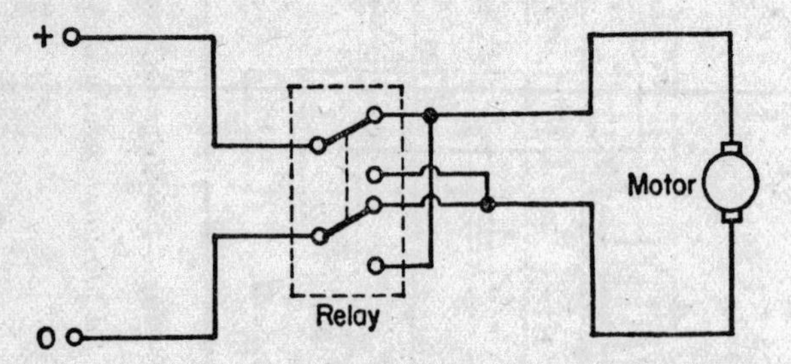

Fig. 11.7 Using a double-pole double-throw relay as a reversing switch.

The system described above allows one to change speed *instantly* from one value to another. Often it is sufficient to increase or decrease speed *gradually* until the required new speed is reached. Since only two commands are required ('increase speed', 'decrease speed') instead of a distinct command for each of the possible values of speed, we are able to effect a considerable economy in the transmission of instructions and can use these savings to provide for instructions of other kinds. This is the approach adopted in the remote control i.c. mentioned using heading (3) in the previous section. Figure 11.8 shows how this function may be added to the digital-to-analogue converter of Figure 11.6. The 74193 is a counter that may be clocked

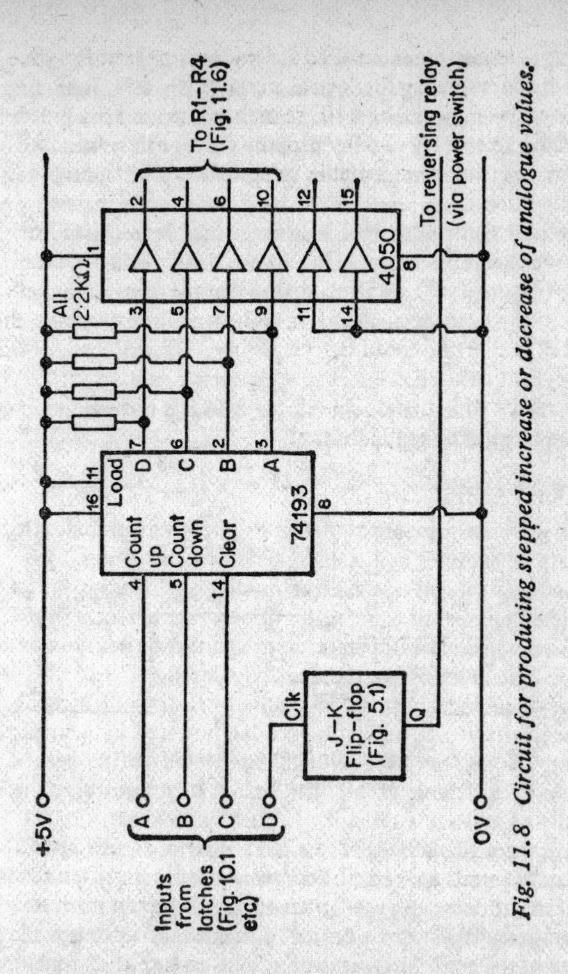

Fig. 11.8 Circuit for producing stepped increase or decrease of analogue values.

to count either up or down from its initial state. A low pulse to pin 4 causes it to increase its count by 1. This takes effect as the input rises from low to high. Count 15 is followed by count 0 when counting up; count 0 is followed by count 15 when counting down. By using this i.c., speed can be controlled over the whole range of the d-to-a converter by using only two lines. The remaining two lines can be used in various ways. Figure 11.8 indicates one possibility. Line C is used to clear the counter by delivering a high pulse, so immediately reducing speed to zero, or to whatever minimum speed has been decided upon. Line D can be used to operate a flip-flop so that the direction of the motor can be reversed by toggle action. On four lines we have complete control of speed and direction, with instantaneous 'stop'. It is also possible to load a pre-set value into the counter. The input pins (see Figure B.2c, Appendix B) can be individually wired either to ground or to +5V so that, when the 'load' input is made low, a predetermined value appears at the outputs of the i.c.

If the 'clear' function shown in Figure 11.8 is not required, this input must be grounded.

Positional control

Ratchet mechanisms are available, or can be constructed, that operate whenever a coil is energised and can be used for moving control surfaces such as rudders and ailerons into a restricted number of different positions. These require only a power transistor to operate them and can be easily controlled by circuits of the types already described. For finer control a servo-mechanism is required. In this the control surface is moved by reduction-gearing powered by a small motor, the surface usually being connected to the gear system by a flexible cable. The gear system also turns the spindle of a variable resistor. In the digital proportional system the variable resistor is wired as part of a pulse generator, and there is a circuit to compare the length of pulse generated locally with the length of pulse received from the transmitter. If they are unequal in length the motor is made to run in the appropriate direction so as to turn the variable resistor and adjust the length of the locally-generated pulse to make it closer in length to the received pulse. When the two pulses match exactly, the motor is stopped, having moved the control surface to the required position.

The digital-to-analogue system operates in a similar way, but instead of matching pulse lengths, we match a current produced by the variable resistor against a current produced by the digital-to-analogue converter. A circuit for doing this is shown in Figure 11.9. The converter output comes from 4 resistors, shown with standard values here, though it could also function with only 3 buffers in action if a 3-digit code is being used. The currents from the converter and from the wiper of the variable resistor VR1 are compared by the operational amplifier. This is a CMOS i.c., so the usual precautions should be taken when handling it (p. 8). This type is chosen because it can work on a low supply voltage and its output swings strongly toward either supply rail. It is half-way between the two rails (= +2.5V) when the two currents are equal, but falls sharply to 0V when the current from the variable resistor is very slightly less than that from the converter. The output from the operational amplifier goes to two pairs of transistors. These are connected as a pair of Schmitt triggers, so that though they may be turned on by an increase in the output voltage from the amplifier, the voltage must fall to a slightly lower level before they are turned on again. This gives a more positive 'snap-action' to the circuit. Note that R12 has value 100kΩ but R13 has the value 10kΩ. The effect of this is that as the voltage from the amplifier rises TR3 turns on slightly before TR1; as voltage falls TR1 turns off slightly before TR3. Thus there is a small range of voltage over which TR3 is on but TR1 is off. The Schmitt trigger circuits act as inverters, so that when TR1 is turned on TR5 is turned off. The three possible states of the circuit are given in the table below, which is useful when checking the operation of the circuit during construction:

Current from VR1 wiper, compared with that from d-to-a converter	*Output of operational amplifier (approx.)*	*State of TR1*	*State of TR3*	*State of TR5 and RLA1 coil*	*State of TR6 and RLA2 coil*
higher	more than 2.8V	on	on	off	off
approx. equal	2.4 – 2.8V	on	off	off	on
lower	less than 2.4V	off	off	on	on

The relay switches can be arranged so that the motor turns one way when the wiper current is greater than the converter current, turns in the opposite direction when wiper current is less than converter current, and is switched off when the two currents are more-or-less equal. By making the connections to the motor of the correct polarity, the direction of turning of the motor can be made such that it turns the wiper of VR1 so as to reduce the difference between the inputs to the amplifier. As the difference approaches zero the motor is switched off and the control surface remains in the corresponding fixed position, until a new command signal is received and the process of finding the balance-point is repeated to obtain a new position of the control surface.

Remotely controlled radio set

As an example of a remote control project for the home, Figure 11.10 shows the circuit for a radio set that can be switched to a number of pre-tuned stations under remote control. The volume can also be changed to a number of pre-set levels. The loudspeaker can be muted, or silenced, instantly and the set returns to the station formerly tuned and to the previous volume level when the 'mute' command is repeated. Provision is made for 4 pre-tuned stations (for

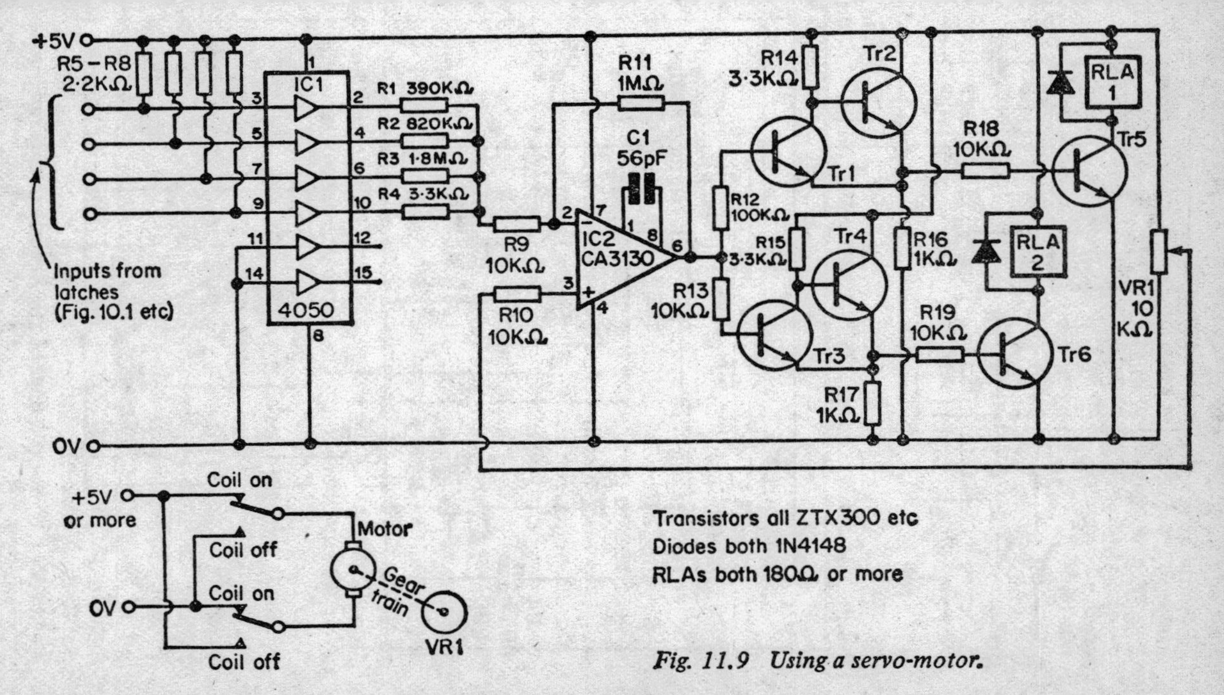

Fig. 11.9 Using a servo-motor.

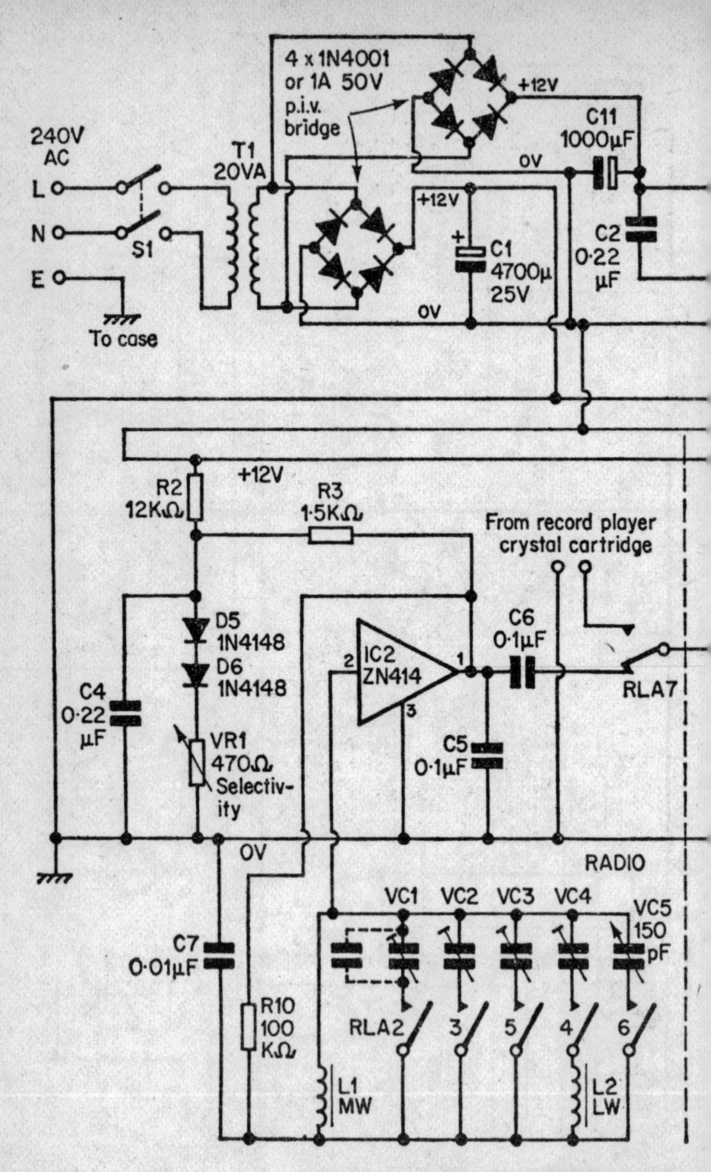

Fig. 11.10 Remote control radio set.

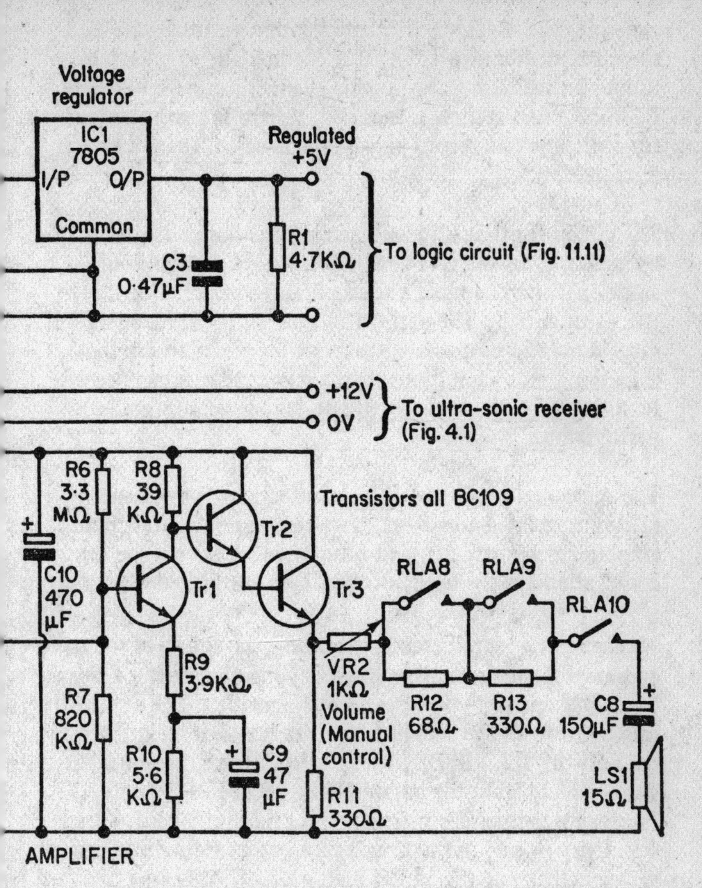

Power supply and radio circuits.

example BBC Radio 1, 2, 3 and 4), one on long-wave and three on medium-wave. A fifth channel allows adjustable tuning on medium-wave, so this channel can be set to, say, the local radio station if required. There is also provision for switching to a record-player crystal cartridge, or a crystal microphone.

The circuit for the radio has been kept as simple as possible by basing it on the Z414 radio i.c. with a minimum of additional components. The amplifier too is a simple three-transistor circuit. The performance of the set is adequate for most domestic purposes. Readers who prefer to attempt something more ambitious may apply similar principles of remote control to a radio circuit that gives top grade performance.

The radio set consists of five main sections: power supply (12V unregulated and 5V regulated), radio detection circuit, amplifier, logic circuits, and ultra-sonic receiver. The set could alternatively be controlled by an infra-red system.

We shall not go into detail in describing the action of the radio and amplifier circuits, for this is beyond the scope of this book, but one or two points need explanation. The automatic-gain-control voltage supplied to the radio i.c. is maintained by the network R2, R3, D5, D6, and VR1. For normal use the voltage supplied to the i.c. should lie between 1.3V and 1.5V. Increasing the voltage increases the gain, but gives lowered selectivity (local loud stations swamp other stations on nearby wavelengths), and introduces distortion. VR1 is used to adjust the a.g.c. voltage, and acts as a selectivity control. Normally it should be set so that selectivity is as high as possible, whilst still providing sufficient gain to receive the desired stations. The preset capacitors VC1–VC4 are switched into the tuning circuit one at a time. In a push-button set, a special mechanical switching system would be used to achieve this, but in this project we operate relay switches by means of logic circuits, as will be described later. When VC1, VC2 or VC3 are switched into circuit, the set is tuned to medium-wave. Each preset capacitor is adjusted to one of the selected stations. If the station is at the high-frequency end of the medium waveband, it can be tuned with the variable capacitor alone. However, stations nearer the low-frequency end of the medium waveband can be tuned only if an additional fixed-value capacitor is wired in parallel with the variable capacitor. The value required can be found by trial: 100pF is usually suitable. When VC4 is switched into circuit this also brings the long-wave tuning coil into circuit. VR5 is a variable tuning capacitor mounted on the panel of the receiver, with a knob for manual tuning. This may be of the air-dielectric type (for example, Jackson Type 0, 1-gang) but smaller, cheaper types can be used if the facility for manual tuning is not regarded as being important. Instead, it is possible to fit another preset capacitor as VC5, and have a fifth pre-selected channel. If the record-player facility is of no interest to you, a sixth pre-selected radio channel can be included.

Construction

The radio circuit is best built on a board separate from other circuits. The whole circuit can easily be accommodated on a small piece of strip-board about 3cm by 6cm. It is important to mount components close together so as to minimise connection lengths. The output decoupling capacitor, C5, should be mounted with its leads as close as possible to the leads of the i.c. The variable capacitors should be connected so that their 'earthy' sides (the terminal connected to the moving vanes) are connected to the relay switches.

The tuning coils are wound on ferrite rods. For the medium-wave coil, L1, use a rod about 12cm long, if possible, though in areas of good reception a rod 6cm long may prove adequate. Wrap a layer of Sellotape around the rod. Then wind a coil of 65 turns, close together so that adjacent turns touch, but do not overlap. The wire used should be enamelled copper wire of 28 or 30 s.w.g. Fix the ends of the coil with Sellotape. When setting up the receiver it may be found that a few extra turns may be needed for best results. It is advisable to wind, say, 70 turns to begin with, for it is easier to remove turns later than to add extra turns. The long-wave coil may be wound on the same ferrite rod (one coil near one end, the other near the other end) but it is preferable to employ a separate rod. For long-wave, use a rod 15 to 20cm long. The coil should have 300 turns of 40 s.w.g. enamelled copper wire. Wind the coil compactly so that it is multi-layered and no more than 1cm long.

The positioning of the various parts of the radio set needs some consideration. The aerial rods must be mounted on brackets made from wood or plastic, and kept well away from metal objects. The case in which they and the radio circuits are housed should be constructed mainly of wood or plastic. The radio-detection circuit should not be located near the loudspeaker, nor near the logic circuit. There is a possibility that radio noise in the form of clicks may be picked up from the logic circuit. To avoid this the logic circuit can be separated from the other circuits by an earthed metal screen. Another approach is to house the power supply and logic circuits in an earthed metal case connected by a multicore lead to a loudspeaker case in which the radio and amplifier circuits are located.

Logical control

We will study the operation of the logic circuit (Figure 11.11) in some detail for it illustrates several general principles that can be applied to the design of other remote control projects. Although this circuit is being used for controlling a set of relays, it could equally well be used for switching on lamps, tone generators or many other kinds of device, as previously described. The main aspects to consider here are the logical schemes used to effect various types of control.

The circuit is driven by an oscillator, or 'clock' running at about 0.6Hz. The pulses from the clock are fed to a decade counter (IC3); the four outputs from this counter run continuously through the sequence 0, 1, 2, . . . , 8, 9 (in binary). The decoder/driver i.c. (IC4) converts this 4-digit binary code to the control signals needed for producing the corresponding decimal digit on the 7-segment LED display. The one used here is of the *common anode* type, and the 150Ω resistor is essential to limit current flow through the LEDs to a safe level.

The input to the circuit comes from an ultra-sonic receiver such as that of Figure 4.1. Since the receiver is operated on a 12V supply and the logic operates on 5V, we must step down the voltage by using a resistor network, as shown in Figure 11.12. The input to the circuit should be normally high, and go low when a command signal is received. This is the case with the ultra-sonic receiver, but should you wish to use this circuit with some other kind of receiver, it may be necessary to use an additional INVERT gate, as indicated in dotted lines, to invert the incoming high pulse. IC1 has a spare NAND gate that can have its inputs wired together to make an INVERT gate. When the command pulse arrives it triggers two pulse generators (IC2). Note that this action is triggered by the leading *edge* of the command pulse. Thus the *length* of the command pulse does not matter. Having an edge-triggered circuit overcomes the problem of what is to happen when the command pulse begins while it is in one state (say, 'switch on Radio 2'), and lasts until the next state ('switch on Radio 3'). Following the instant at which the leading edge of the control pulse triggers the pulse generators, the long and short pulses from the generators take over control of the circuit. The operation of the three groups of functions will be considered separately, since each illustrates a different principle of control:

(1) *Volume:* The initial volume level is set by adjusting VR2 manually; this sets the highest volume level required. The volume can then be reduced in three stages by remote control. A pulse at stage 8 reduces volume slightly; a pulse at stage 9 reduces it by a greater amount; a pulse at both stage 8 and stage 9 reduces it to mere background level. To restore volume to its maximum level a command pulse is sent at stage 1. Figure 11.11 shows that this part of the logic circuit consists of two bistables, set by a low pulse from outputs 8 and 9 respectively of IC7. They are reset by a low pulse from output 1. To understand the action of IC7 we must go

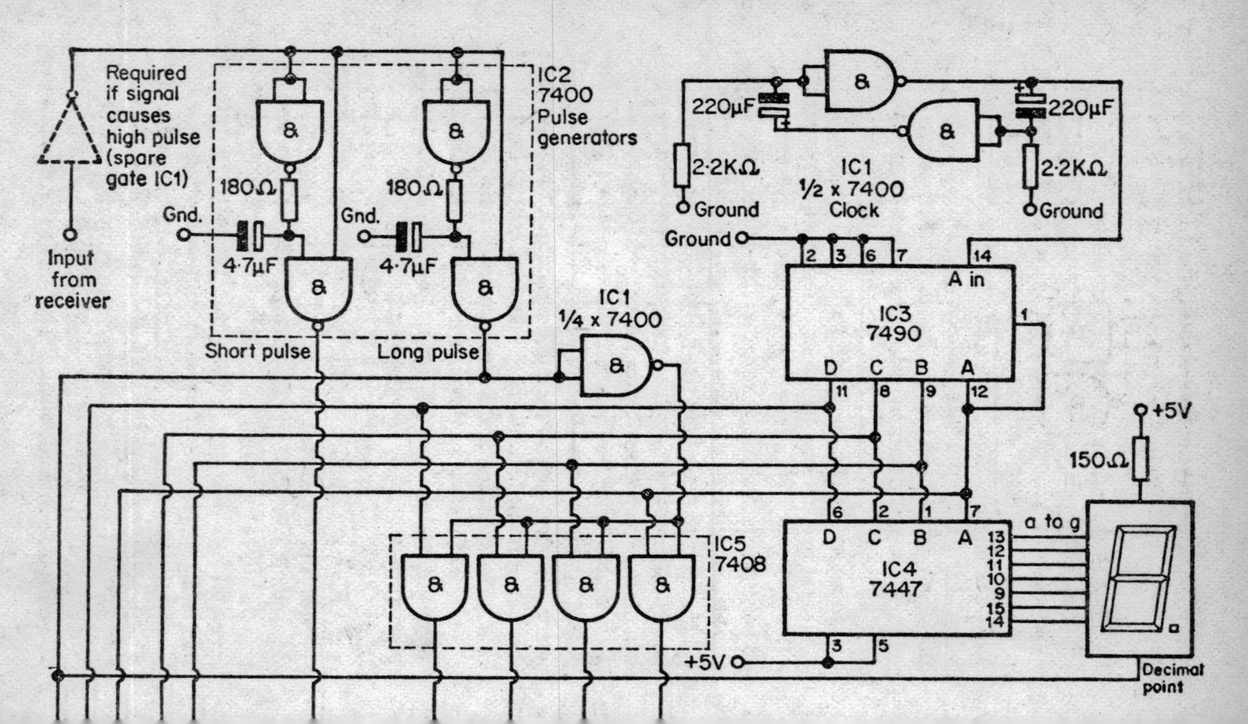
Required
if signal
causes
high pulse
(spare
gate IC1)
Input
from
receiver
IC2
7400
Pulse
generators
180Ω
180Ω
Gnd.
Gnd.
4·7µF
4·7µF
Short pulse
Long pulse
IC1
1/4 x 7400
220µF
220µF
2·2KΩ
2·2KΩ
Ground
Ground
IC1
1/2 x 7400
Clock
Ground
A in
IC3
7490
D C B A
IC4
7447
IC5
7408
+5V
+5V
150Ω
a to g
Decimal
point

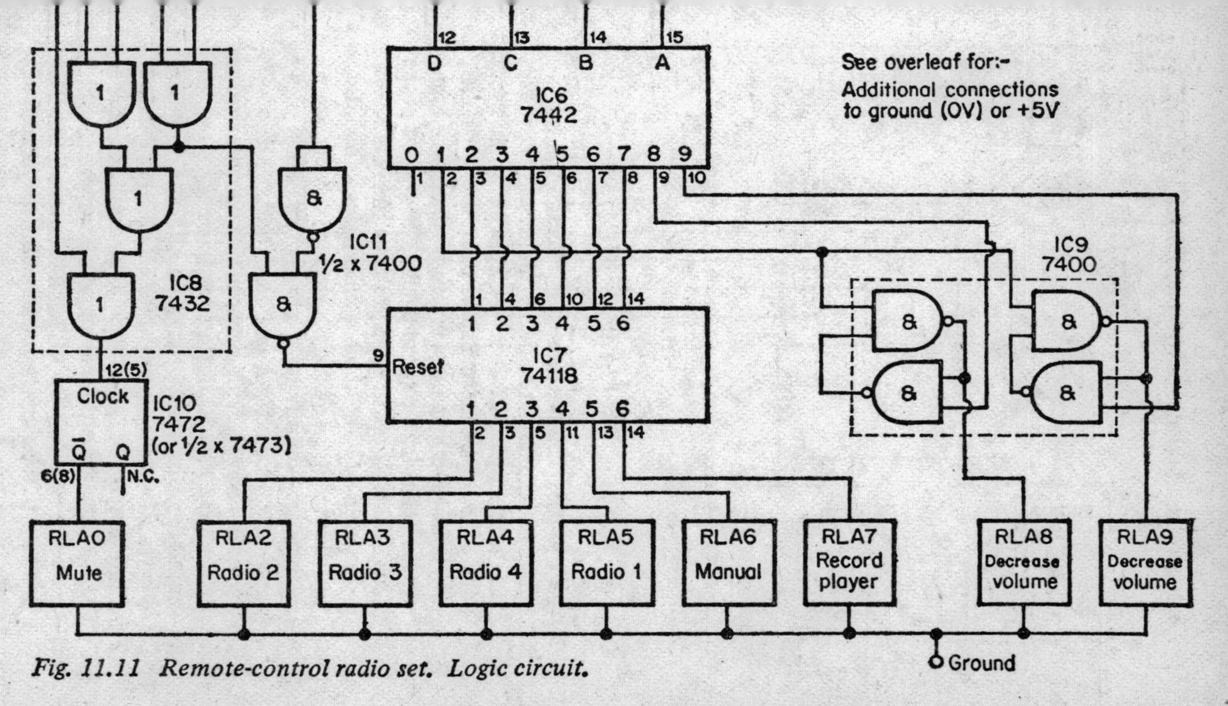

Fig. 11.11 Remote-control radio set. Logic circuit.

ADDITIONAL CONNECTIONS TO GROUND (OV) OR +5V

Pins	1	2	3	4	5	6	7	8	9	10	11	12	13	14	15	16	
IC1							0							5	-	-	
IC2							0							5	-	-	
IC3		0	0		5	0	0			0					-	-	
IC4			5		5			0								5	
IC5							0							5	-	-	
IC6								0								5	
IC7								0								5	
IC8							0							5	-	-	
IC9							0							5	-	-	
IC10		⑤	⑤	⑤	⑤		0		⑤	⑤	⑤		⑤	5	-	-	(7472)
IC10	⑤	⑤	⑤	5		⑤	⑤			⑤	0			⑤	-	-	(7473)
IC11							0							5	-	-	

⑤ = connect to +5V through 1KΩ resistor.
Also connect *unused* inputs
of IC1 and IC11 in this way.

Fig.11.11 (continued)

back to IC3. This has four output lines, A to D, representing the stage of the circuit in binary code. In IC5 these outputs are each ANDed with the inverted long pulse. The long and short pulses are low pulses, so the inverted long pulse is a high pulse. When this arrives (following a command pulse) the outputs of the four AND gates of IC5 show the same binary numbers as is shown by IC4. At other times outputs from IC4 are all low. IC6 is a binary-to-decimal decoder. Its outputs are normally high, but any *one* output goes low when the corresponding binary number is fed to its inputs. Suppose, for example, that the outputs from IC3 at stage 9 are 1001; *if* a command pulse is received at this stage, inputs 1001 are fed to IC6 and its output 9 gives a brief low pulse, which sets the bistable and turns on relay 9. All other outputs from IC6 remain high, so the rest of the circuit is unaffected.

(2) *Station:* The operation of the 'station-select' function is similar to the above, but different in the method of

resetting. The outputs 2 to 7 from IC6 go to IC7 which contains six bistables, like those used for controlling Relays 8 and 9. The bistables are all reset together by a low pulse to pin 9 of IC7. The difference in operation in this function is that in volume control we used the action of resetting all (two) relays to obtain a useful function – maximum volume. To reset all station selecting relays leaves the tuning circuit as an open loop that tunes to several stations at once. We must ensure that this does not happen; one relay from the group 2 to 7 must *always* be energised. Actually this is not strictly possible, for all bistables *must* be reset together, so it is arranged that when a station change is required, *all* bistables are reset and the new station is immediately selected. This is the idea behind the short and long pulses. The bistables are reset by the short pulse; then the long pulse (via IC5 and IC6) sets the bistable corresponding to the newly-required station. This action is further complicated by the fact that the short pulse is produced at every command pulse, so all stations might be reset when we were commanding an alteration in volume level. We need to pick out which commands refer to station selection and which refer to volume control or muting. This is where a certain amount of compromise is needed. Originally the intention was that stages 1, 2, 3 and 4 would select Radio 1, 2, 3 and 4 respectively. However, it would need some rather complex gating to decode the outputs of IC3 so as to distinguish stages 1–6 from stages 7–0. The solution is much simpler if we use stages 2–7 for station selection. Each of stages 0, 1, 8 and 9 in binary has zeros for the middle two digits (0000, 0001, 1000 and 1001), whereas the binary numbers 2 to 7 all have at least one '1' in the middle two digits. The gating for detecting two middle zeros is simple, we just OR the B and C outputs from IC3. This operation is also required for the muting control, so no additional gate is needed. At stages 2 to 7 the output of the OR gate in IC8 is high; this is NANDed with the inverted short pulse giving a short low pulse to the rest of IC7 if a command pulse is received during stages 2 to 7 *only*. This is a good example of how an ideal design may be modified to achieve reduction in circuit complexity. Since many listeners will not want to tune all four BBC stations, it is not worth while to make the logic unnecessarily elaborate.

(3) *Muting:* Here we need a flip-flop action; at the first command the set is muted; at the next command the previous station and volume level are restored. This is obtained by using a flip-flop i.c., the 7472 (IC10). If you were designing for another application and needed another flip-flop, the 7473, which contains two flip-flops would be more economical to use. The flip-flop is to be operated only during stage 0. This stage is detected by using OR gates (IC8). Only when all four outputs of IC3 *and* when the long pulse is triggered does the output to the clock input of IC10 go low, to cause the flip-flop to change state.

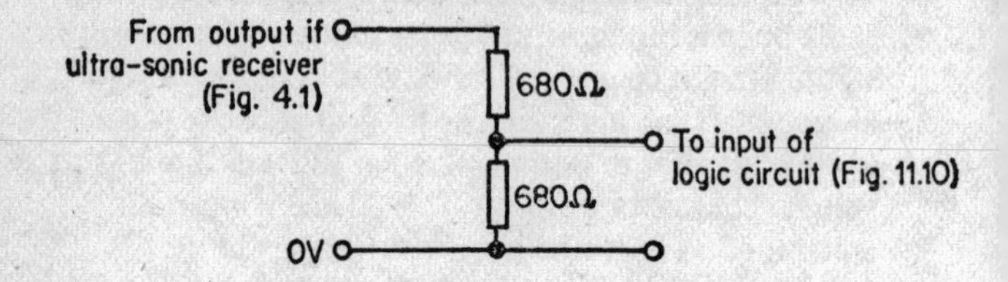

Fig. 11.12 Interfacing ultrasonic receiver to the remote-control radio.

In Figure 11.11 the relays are shown connected directly to the outputs of the i.c.s. This is allowable if the correct type of relay is used, such as the d.i.l. reed relay designed for TTL operation. Other types of reed relay are much cheaper but may require transistor drive. There is some advantage in this, since not only does this place less load on the TTL i.c.s, but it also allows the relays to be powered directly from the 12V line, so reducing the load on the 5V regulator i.c. The circuit for relay-driving from TTL is given in Figure 11.13. If you are using a relay intended for operating on 6–9V, wire a 470Ω resistor in series with the relay coil. Since there are several relays to drive, you can save time and space by using a transistor array (e.g. CA3046) instead of separate transistors. Ideal in this respect is the darlington driver i.c. which incorporates the input resistors, the switching circuits and the protection diode for seven separate relays (or small motors, lamps, etc.) at considerable saving of cost in addition to the convenience of the great simplification of wiring.

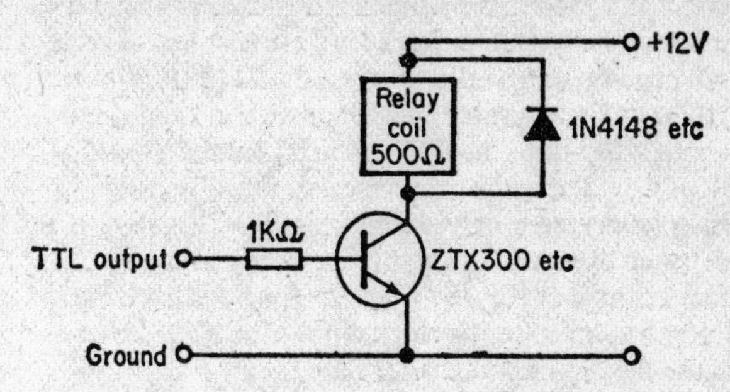

Fig. 11.13 Transistor drive for relay coil.

12 DUAL CONTROL

Sooner or later you may want to be able to control two or more devices simultaneously. It is essential that the control signals from one system are not picked up and acted upon by the other system or systems. With radio control it is usual to operate each system on a different radio frequency. Transmitters and receivers are both accurately tuned, using calibrated crystal oscillators, so that there is no chance of interference between adjacent wavebands. When we are using other methods of transmission, such as infra-red, for example, we need to introduce something analogous to radio-frequency. To achieve this we transmit a *tone burst* (Figure 12.1) instead of a simple pulse. The 'pulse' now consists of a series of pulses at high frequency. The receiver is designed so that it responds only to tone bursts of the correct frequency. In other words, the receiver is tuned to the frequency of the transmitter. It is then possible to have pairs of transmitters and receivers tuned to different frequencies, and each pair can operate without interference from others in the same area. It is also possible for one transmitter to be switchable from one frequency to another so that we can control several devices independently from a single transmitter.

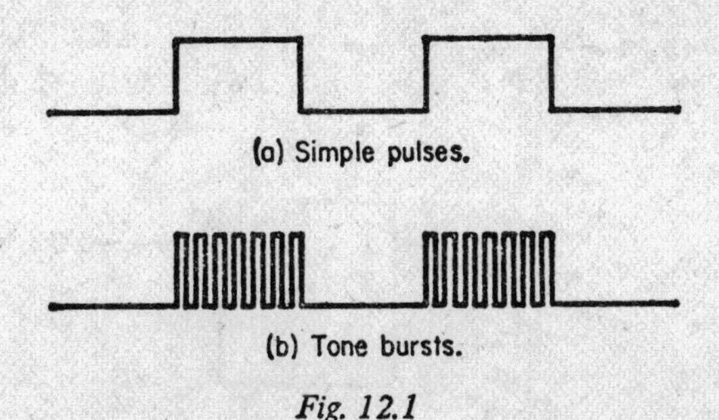

Fig. 12.1

The use of tone bursts instead of simple pulses also confers the advantage that there is greater freedom from interference from external sources. An infra-red system or a visible-light system may be prone to interference from sources of radiation such as room lights, heaters or the sun. If receivers are designed to respond only to tone bursts, external sources are ignored.

Transmitting tone bursts

The most convenient source for the high frequency is an astable multivibrator built from a 555 time i.c., as in Figure 12.2. The variable resistor is used for adjusting the frequency. With the values given, frequency can be adjusted over the range 300Hz to 10kHz. When a suitable frequency has been found, VR1 and R1 could be replaced by a single fixed resistor of the required value. The circuit could also be modified to allow switching of resistor value between the 5V line and pin 7, so as to allow the frequency to be set instantly to various values for controlling several devices from the same transmitter. An alternative way of providing a range of frequencies is to use one or more flip-flops to divide the frequency by 2 at each stage. A circuit for frequency division is shown in Figure 5.2. If the output from the 555 is fed to the input of this circuit, a waveform of half the frequency of the 555 is obtained from Output 1, and a waveform of one quarter the frequency is obtained from Output 2.

The output from the 555 oscillator, or from any flip-flops connected to it, can be used directly to power an infra-red

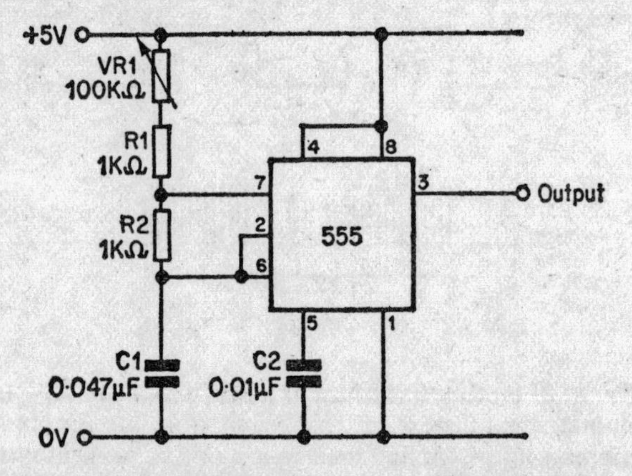

Fig. 12.2 Astable multivibrator using 555 timer IC.

LED, as in Figure 12.3a. To drive a low-voltage filament lamp, use a transistor, as in Figure 12.3b. This allows transmission over greater distances (Chapter 6). The push-buttons are intended as 'transmit' buttons. When they are pressed a tone burst of infra-red radiation or visible light is transmitted. This is how these circuits are used in a single-pulse system or for sequential control (Chapter 5). For use in multiple-pulse systems the circuits of Figure 12.3a and b are operated by logic gates. In the case of the multiple-pulse coder (Figure 8.4) the output from the 555 must be ANDed with the output from the shift register of the coder. This may be done either with an AND gate or with two NAND gates, as shown in Figure 12.3c.

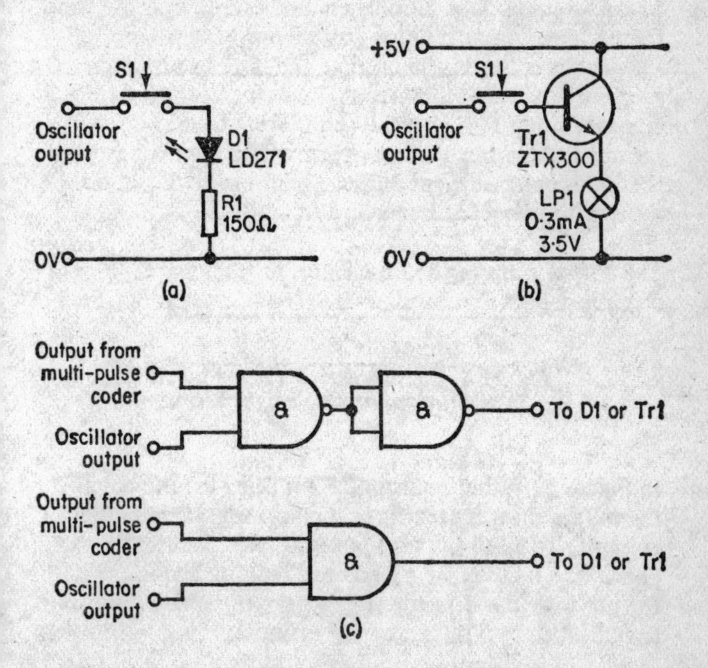

Fig. 12.3 Producing tone bursts from an LED or filament lamp.

Detecting tone burst signals

By using the circuit of Figure 12.4 the signal can readily be monitored by ear, assuming that the tone is in the audio-frequency range. The values used in Figure 12.2 provide a range of tones, all of which are audible. This circuit is very useful when experimenting, and when setting up and testing an infra-red or visible-light link. Incidentally, the circuits of Figures 12.3a or b and 12.4 together make up a simple transmitter-receiver system for morse-code signalling by light-beam. Since this does not operate on radio-frequencies it is an entirely legal method of transmission and no licence is required. If, instead of using the 555 i.c., you take the output from an audio-amplifier, this system can be used to transmit speech or music.

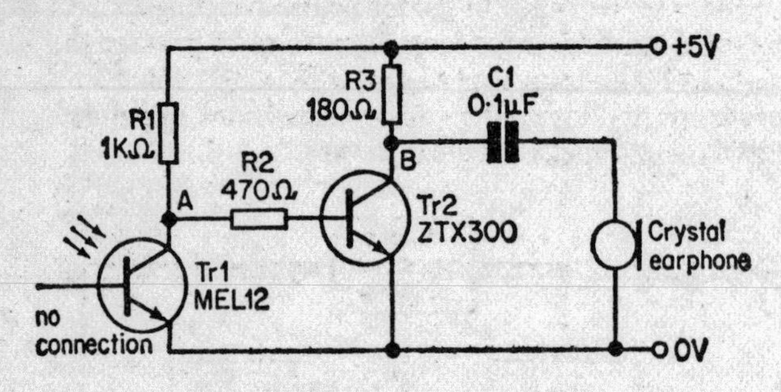

Fig. 12.4 Simple receiver for ton-bursts in an infra-red or visible light beam.

In Figure 12.4, the transistor TR2 is being used to amplify the voltage changes occurring at point A when a signal is being received. To make an even simpler receiver, connect the earphone directly to point A, and omit TR2, R2 and R3. But for use with the detector circuit next to be described, amplification by TR2 is normally essential.

The essential features of a circuit that detects tone bursts of a particular frequency are shown in Figure 12.5. A voltage-controlled oscillator (VCO) is set to oscillate at a given frequency, the central frequency, which is the frequency to which we wish the circuit to respond. The frequency of the VCO is varied on either side of central frequency by varying the voltage applied to it from the loop filter. The phase comparator compares the input signal (which may come from a circuit like that of Figure 12.4) with the oscillations from the VCO. The output of the phase comparator is proportional to the error between the phase of the input signal and the phase of the signal from the VCO. This output is filtered by the loop filter. The varying voltage from the loop filter controls the frequency of the VCO making it change until it locks on to the frequency of the input signal. Thus the complete circuit, which is a closed loop, is known as a *phase-locked loop* (PLL). Output from the loop may be taken at several different stages, depending upon what functions are required. In the tone-burst-detecting application, output is taken from the loop filter, so as to monitor the control voltages being applied to the VCO. To minimise the loading on the loop filter, this voltage is not monitored directly, but through a source follower.

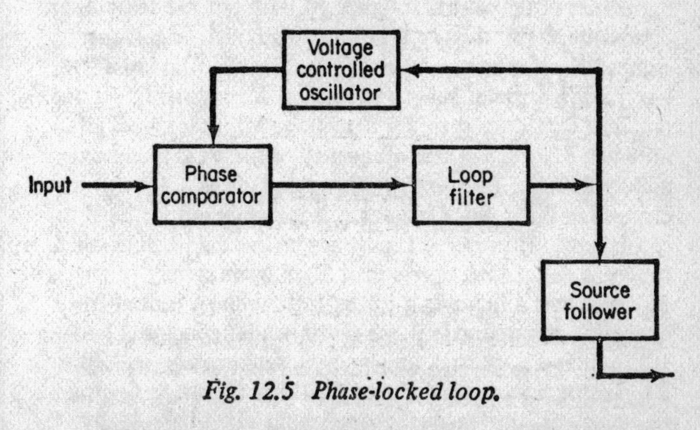

Fig. 12.5 Phase-locked loop.

The complete circuit for a phase-locked loop is obtainable in a single CMOS i.c., the 4046. It needs only a few external capacitors and resistors. The i.c. contains two sections of the circuit as separate units: the VCO with source follower, and the phase comparator. Actually this i.c. contains *two* phase comparators, each with distinctive operating features. Phase

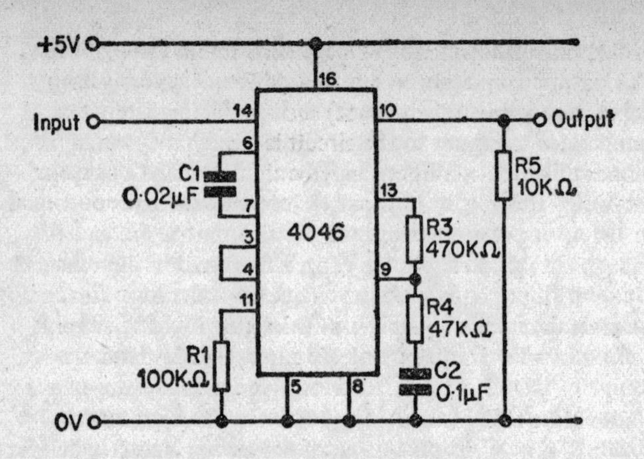

Fig. 12.6 Practical phase-locked loop circuit.

comparator I is a low-noise phase detector (simply an exclusive OR gate), with a relatively narrow tracking range. It requires a 50% duty cycle for the input signal and also locks on to harmonics of the central frequency. Both of these features are disadvantages in this application. Although the Phase comparator II is more subject to interference from noise it has the advantage of being a wide-band detector and does not require a 50% duty cycle for detection. With the values given in Figure 12.6, the central frequency of the VCO is approximately 300Hz. To work at other frequencies, C1 may be changed in value, since frequency is inversely proportional to capacitance. However, C1 must not be less than 100pF when supply voltage is 5V, or less than 50pF when supply voltage is 10V and over. Increasing R1 to 1MΩ reduces central frequency to approximately one-tenth, while reducing R1 to 10kΩ increases central frequency approximately tenfold. R1 must not be outside the range 10kΩ to 1MΩ. Central frequency is also increased approximately fifty-fold by increasing supply voltage to 10V or over. The loop filter consists of R3 and R4 with C2. With the values shown the circuit works as described and there is no need to alter these values.

The output from the source follower (pin 10) must be fed to a load resistor R5, which must have a minimum value of 10kΩ. The output may then be fed to a device of suitably high impedance (greater than 10kΩ) such as the input of a CMOS gate. When the input to the circuit is steady at 0V or at 5V (either of which conditions may be thought of as a signal of frequency 0Hz), output at pin 10 is low (0V). It remains close to 0V at input frequencies below the central frequency, though rises gradually as the central frequency is approached. When the input equals central frequency, the output reaches its maximum value, which is a little over 3V, at a supply voltage of 5V. This is effectively a logical high and can be fed to a CMOS logic gate. Increasing input frequency above central frequency has no further effect; output remains steadily high. Thus the output of the circuit is high whenever it is receiving a tone-burst or frequency equal to or higher than the central frequency.

One way of operating two remotely controlled devices independently from two transmitters or from a single switched-frequency transmitter is shown in Figure 12.7. Device 1 responds when the tone burst frequency is 800Hz because its phase-locked loop is tuned to that frequency. Device 2 responds when the tone-burst frequency is 400Hz, obtained in this example by dividing the timer frequency by 2, using a single flip-flop. Device 2 requires two phase-locked loops, one tuned to 400Hz and the other tuned to 800Hz. Their outputs are fed to an exclusive-OR gate. When the received tone-burst has frequency 800Hz, both loops respond, both outputs are high and the output of the gate is low. When there is no tone-burst, both outputs are low, and the output of the gate is low. But when the tone-burst frequency is 400Hz, the outputs of one loop (400Hz) is high and the other one (800Hz) is low; this is the exclusive-OR condition, and the output of the gate goes high, thus passing the pulse to the decoder.

In this chapter we have touched upon the application of phase-locked loops to remote control. The design and application of phase-locked loops is a subject in its own right, and for further information on using these devices the reader is referred to the specialist books.

Another method of controlling two or more devices from a single transmitter or for operating two or more pairs of transmitters and receivers in the same area is described in Chapter 13.

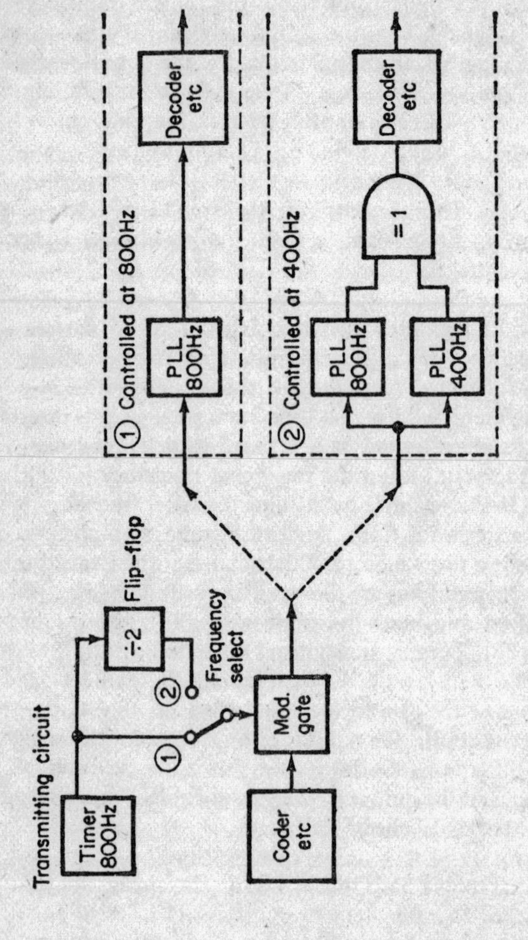

Fig. 12.7 Dual control from a single transmitter.

13 PULSE POSITION MODULATION

There are four main methods of conveying command signals:

(1) *Single pulse:* produces a single response, though by suitable receiving circuits it may be made to control a number of functions in a sequential manner (Chapters 3, 4 and 5).

(2) *Multiple-pulse:* pulses of standard length; the *value* of the pulse (0 or 1) conveys the information (Chapters 8, 9 and 10).

(3) *Digital proportional control:* pulses are sent at regular intervals but the length of the pulse is varied. Pulses may be of two kinds, 'short' and 'long', the 'long' usually being twice the length of the 'short'. The message is interpreted as a series of 1's and 0's and decoded as a binary code. Analogue information can also be transmitted, the length of the pulse being varied over a given range in proportion to the value of the analogue quantity. Digital proportional control is widely used in commercially-built radio-control systems, and since articles on such systems for home construction are frequently published in the hobby electronics press, this type of control will not be discussed further here.

(4) *Pulse position modulation:* pulses are all of equal length. The *intervals* between pulses are either 'long' or 'short', giving a binary-coded message.

The waveforms of the systems are compared in Figure 13.1. In all systems the waveforms may be used to modulate a carrier-wave. For example, the single-pulse ultra-sonic transmitter emits a fully modulated wave (either full on or full off) at 40kHz. Visible light and infra-red transmission usually do not employ modulation unless a frequency-sensitive circuit is being used as a tone detector to overcome interference (Chapter 12). With radio control, the 27MHz carrier radio wave is fully modulated (Chapter 15).

Single-pulse circuits and multiple-pulse circuits are relatively

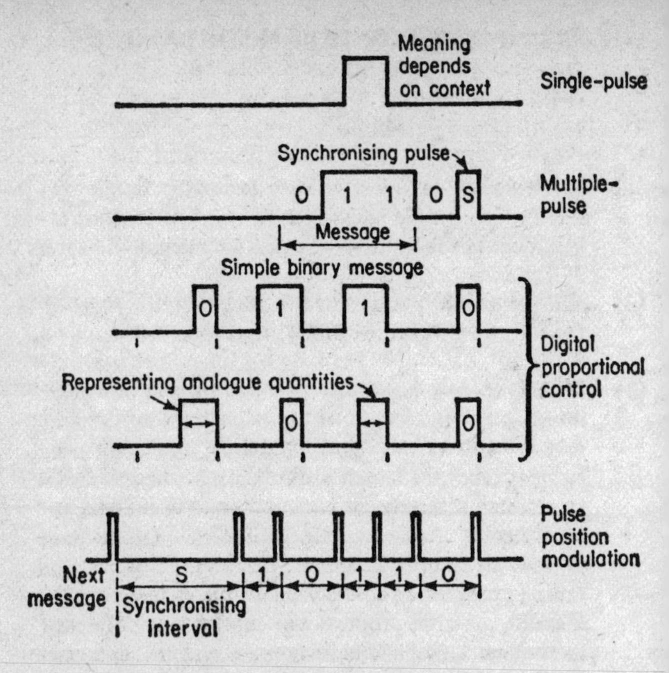

Fig. 13.1 Waveforms of various types of control signal.

easy to design and build. Digital proportional control is generally more complex, especially if many channels are required. For pulse position modulation, a pair of specially designed integrated circuits is available.

The 490 coder i.c.

This i.c. can accept up to 21 different commands. A command is given by connecting one of eight *current source* pins to one of four *current sink* pins. The latter include pin 1 which is grounded. Only the connections indicated by circles in Figure 13.2 give meaningful commands. The commands may be thought of as 5-digit binary code groups, having the format EDCBA. Figure 13.2 is marked to show how each group may be coded by striking the appropriate key. There is no need to fit more keys than are actually required to give the commands you need. The commands are as follows:

(1) *Program codes* (0000X to 1001X): In these 10 codes, digit A is ignored by the coder. The four digits E to B can thus be thought of as a 4-digit binary number, coding programs 0 to 9.

(2) *Program step* (10101, 11101): These cause the program outputs of the decoder i.c. to step through their sequence 0 to 9 forward or backward, respectively. The rate of stepping is controlled by a timing circuit at the decoder.

(3) *Analogue increase* (10100, 10110, 10111): The decoder i.c. has 3 outputs, 'Analogue 1', 'Analogue 2', and 'Analogue 3'. These are useful for the control of motor speeds and other analogue functions. For the reasons given on p. 77, these are not true analogue outputs, for they increase by steps, but since there are 32 steps in the range of each output the degree of control is sufficiently fine for most purposes. Pressing and holding an analogue increase key causes the analogue output to be stepped up at a predetermined rate until it reaches its maximum value, after which there is no further increase. The outputs come from *current* regulators; a reference current can be set at the decoder i.c. and this determines the level of currents supplied from the analogue outputs.

(4) *Analogue decrease* (11100, 11110, 11111): These commands make the analogue outputs step down, continuing until zero current output is reached.

(5) *Standby* (11000): This controls the 'On/Standby' output of the decoder i.c. When the i.c. is first switched on, this output is high. When a program code or program step command is received, it goes low. It goes high again when a standby command is received.

(6) *Toggle output* (11001): Controls the 'toggle' output on the decoder i.c. This is initially low, but changes to high when Analogue 2 output is brought to zero level.

(7) *Normalise* (11011): At this command, all analogue outputs are taken to 12/8 of reference current, and the 'toggle' output is taken to low.

From the description above it is clear that the range of control functions that may be achieved by this pair of i.c.s is enormous

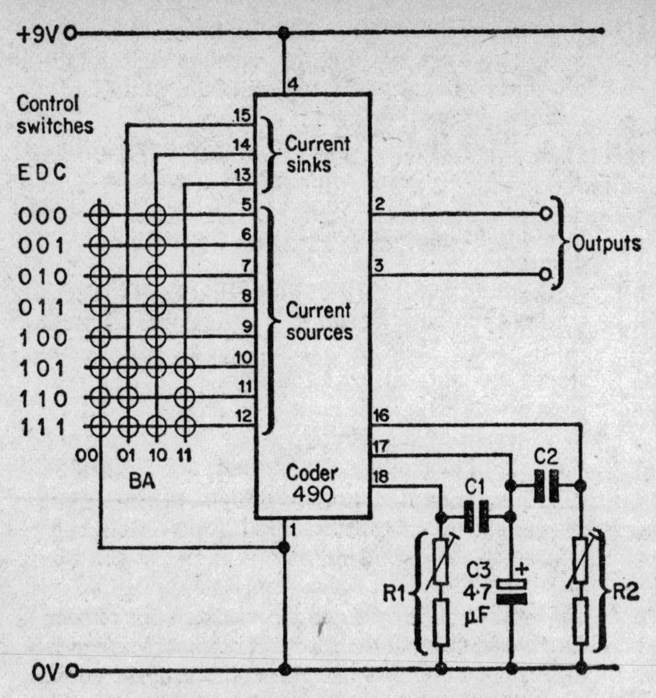

Fig. 13.2 Basic circuit for coder IC.

and limited principally by the imagination and ingenuity of the user. In planning a particular project, it becomes a fascinating exercise to make the very best of the facilities offered. A relatively simple example is given later (Chapter 16).

Frequency setting

If transmission is to be by ultra-sound, or by tone bursts of visible light or infra-red, or if a modulated signal is to be sent by cable, a *carrier frequency* must be generated. The circuits for doing this are contained within the coder i.c. and require only a timing capacitor C1 and resistor R1. Figure 13.2 shows that R1 should consist of a fixed resistor plus a variable preset resistor so that the frequency may be set to the desired value. The equation for calculating frequency is:

$$f \approx \frac{1}{C_1 R_1}$$

where f is in hertz, C_1 in farads and R_1 in ohms. The resistor should have a value between 20kΩ and 80kΩ, and f may take any value up to and including 200kHz. If no carrier frequency is required, as in *un*modulated infra-red transmission, omit C1 and wire a fixed resistor, value 2.2kΩ for R1.

The other frequency that must be set is the *modulation rate.* This determines t_0, the time for a '0' interval. The '1' interval then has 2/3 of this value, and the 'S' interval is twice the '0' interval. The equation for calculating modulation rate is:

$$t_0 \approx 1.4 C_2 R_2$$

where t_0 is in seconds, C_2 is in farads, and R_2 is in ohms. The resistor should have a value between 15kΩ and 100kΩ, and t_0 can lie between 1s and 0.1ms.

Choice of carrier frequency depends upon the mode of transmission. For most ultra-sonic crystals, the resonant frequency is approximately 40kHz and the values of C1 and R1 must be chosen so as to obtain this. Choice of modulation rate is affected mainly by the length of time required for the stepping of the analogue outputs through their full range. Although the time required to transmit and receive a command may be a matter of only a few hundred millisecond, the analogue outputs have 32 steps and the stepping command must be transmitted and received for each of these steps. In practice each command must be transmitted and received *twice* before it is effected, since the decoder has an error-checking feature that requires it to receive the same code twice in succession before it responds. This means that the time taken to step from one end of the scale to the other may exceed 10s even when t_0 is only 27ms, a bit rate of 37/s. If an ultra-sonic transmitter is being used it is not possible to decrease t_0 to less than 13ms without some loss of range, for the transducer rise-time is of the order of 2ms. Using infra-red and radio, t_0 may be as little as the minimum, 0.1ms.

Outputs

When a key is pressed the corresponding 5-bit code is transmitted, followed by the synchronising 'S' bit, and this sequence is repeated for as long as the key is held down. When the key is released the coder continues to the end of a code group and then stops. There are two output pins, each producing pulses in antiphase (Figure 13.3), modulated or not, depending on whether C1 is included in the circuit.

Fig. 13.3 Output waveforms.

A further output is available from pin 17. This is normally low, but goes high whenever a key is pressed. This output is useful, for it can operate an LED to give indication that a signal is being transmitted. Excessive current drain must be avoided, so this indicator must be driven by a transistor, as shown in Figure 13.4a. A simpler method of indicating transmitting action may be employed if output 2 is not required for activating the transducer. As figure 13.3 shows,

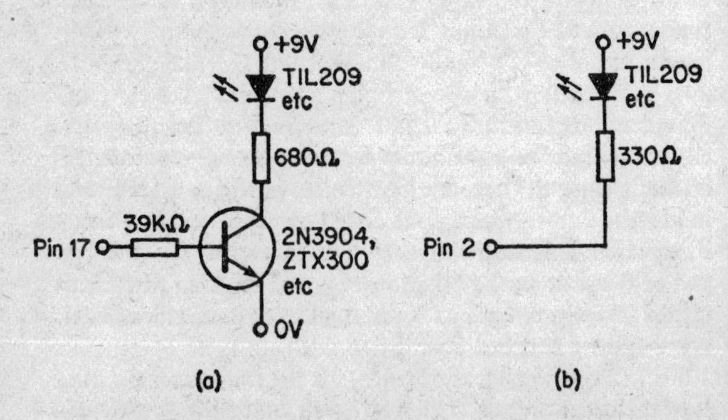

Fig. 13.4 Providing visual indication of transmission.

the output of pin 2 is high when pulses are *not* being transmitted. An LED wired as in Figure 13.4b is dark between pulses but flashes when a code group is being transmitted.

The way the outputs are used for transmission depends on the kind of transducer involved. For ultra-sonic control the transducer may be connected directly to pins 2 and 3. These provide a current up to 5 mA which is sufficient to give the transmitter a range of up to 8m. For greater range, up to 10m, extra power may be gained by using outputs 2 and 3 to drive transistors, as in Figure 13.5, so as to pull the voltage on one line almost to +9V when the voltage on the other line is zero.

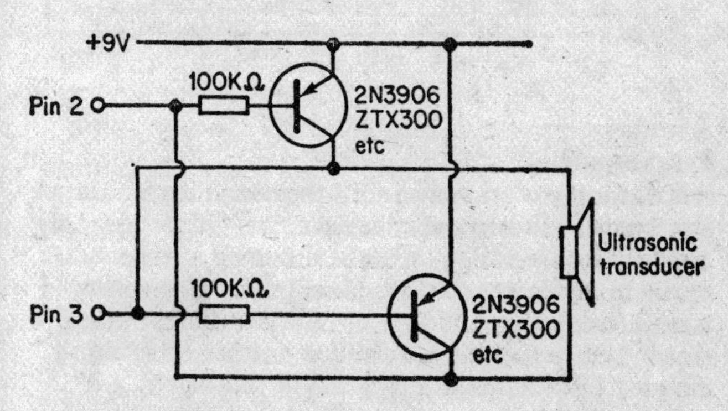

Fig. 13.5 Increasing the range of an ultrasonic transmitter.

The current available from pins 2 and 3 is insufficient to drive a high-power infra-red LED, so this (or *these* if greater range is required) must be driven by transistors. A circuit is given in Figure 13.6 for high power LED transmission. This uses pin 3, which has low output between pulses, leaving pin 2 free to power a visible light LED as indicator (Figure 13.4b). This circuit can be used either with modulated transmission or unmodulated transmission.

Methods for using this i.c. with radio control systems are dealt with in Chapter 15.

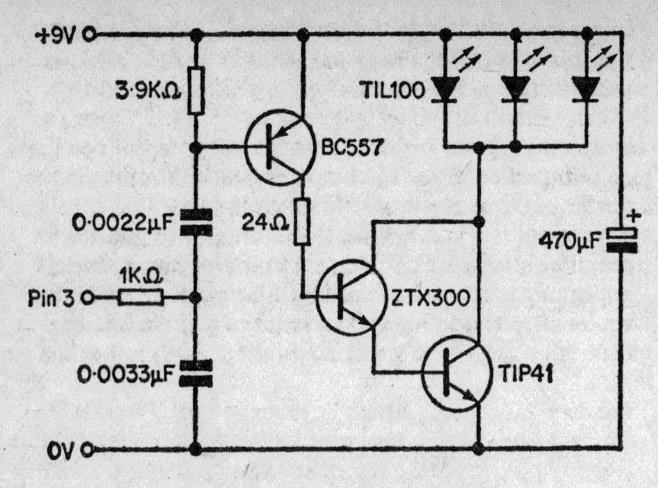

Fig. 13.6 Driving infra-red LEDs.

Power supply

Whereas many other systems of transmission require several i.c.s, resulting in heavy current drain, the 490 requires only 6μA while on standby. While transmitting, the power requirement depends on the mode of transmission employed and whether or not an indicator LED is included in the circuit. For ultra-sonic transmission or when using a single infra-red LED, sufficient power can be supplied by a PP3 battery. This is very convenient if a small hand-held transmitting unit is to be built.

Control panel

In the P.P.M. system all commands are made by pressing keys or operating switches. There is no convenient way in which the joystick control lever of the digital proportional system with its two potentiometers can be coupled to the system. The function of the joystick or steering-wheel control is taken over by pairs of keys. One advantage is the reduction in cost; another is the long-lasting nature of key contacts compared with the tendency to wear shown by potentiometer tracks. Users of joysticks may at first find the unfamiliarity of key-control a drawback, but one soon becomes accustomed to the new method. As an alternative to a pair of keys for operations such as 'left–right', 'forward–backward' or 'climb–dive', a single-pole changeover toggle switch, with a central 'off' position may be used. These can be obtained spring-loaded so that the switch returns to its central position in the absence of pressure. These switches can be mounted on the control panel, orientated so that the direction of movement of the switch lever corresponds with directional commands. If desired, a piece of plastic or metal tubing can be attached to the lever to lengthen it, so that it almost resembles a joystick in action and appearance. Again there is a considerable saving in cost.

For digital functions, such as Program Control, Program Step, Normalise etc., simple key-board switches can be used. Before laying out the keyboard give thought to the frequency with which certain keys will need to be operated, and their sequence of operation, so that the keyboard will be convenient to handle. Apart from the keys, the only other item on the control panel is the indicator LED and, possibly another LED connected so as to indicate that power is switched on.

Dual control

The coder operates at a preset modulation rate and, as will be described later, the decoder i.c. responds only to signals that have the correct modulation rate. It is therefore a simple matter to operate two or more pairs of transmitters and receivers in the same area without any danger of interference. The only concern is to set the modulation rates so that they are different by a sufficient factor. Similarly, a single coder i.c. can be wired as in Figure 13.7 so that its modulation rate can be switched to one of two (or more) values. Two (or more) decoder can then be tuned each to respond to a different frequency, giving independent control of two devices from a single transmitter.

Tuning the coder and decoder

Ideally this should be done with the help of an oscilloscope. The output from the coder i.c. is monitored so that the length of t_0 can be adjusted to the required value. If the transmitter

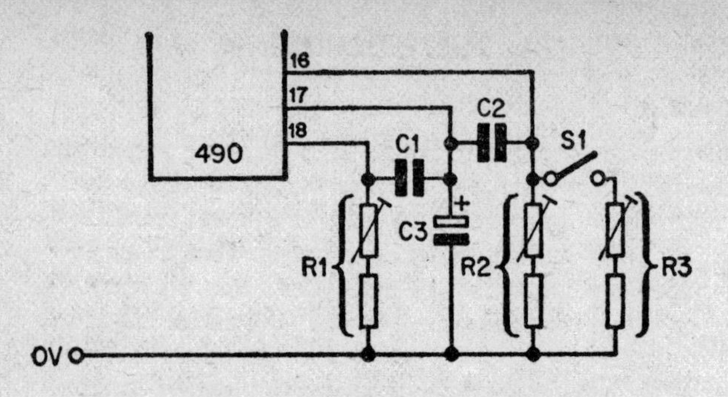

Fig. 13.7 Switching resistor chains in parallel to give two different modulation rates.

is to be used with an ultra-sonic crystal transducer, the carrier frequency is also monitored and adjusted to 40kHz, or whatever the resonant frequency of the crystal may be. The transmitter and receiver are then placed a few metres apart and a transmitter key held down, so as to transmit any one of the program signals 1–9 continuously. The receiver circuit may then be adjusted for maximum response. It may also be necessary to readjust carrier frequency in the transmitter to obtain maximum response in the receiver. The oscillator frequency (pin 6) should also be monitored and adjusted until its period is 1/40th of that of t_0.

Those without an oscilloscope may find that it takes a little more time and trouble to obtain good transmission and reception but the task is not too difficult. One has to rely on the component values at transmitter being within a small tolerance range, otherwise it may not be possible to adjust oscillator frequency to 1/40 of t_0. The use of close-tolerance components for R2 and C2 in the transmitter and for C1 and R2 in the receiver will help reduce the element of uncertainty. With the transmitter in continuous operation *slowly* adjust VR1 until a response is obtained at the receiver. Remember that a response may take a second or two to appear if reception is bad.

14 P.P.M. DECODING

The coder i.c. described in the previous chapter is paired with a special decoder i.c. that decodes the P.P.M. signal and produces the corresponding outputs. One great advantage of the system is that the decoder i.c. is triggered by the *leading edge* of each pulse, not by the pulse itself. This is a consequence of pulse position modulation. The actual *length* of each pulse does not matter (provided it is less than the time-interval between pulses), thus giving this system great immunity from noise. In multiple-pulse systems, and the digital proportional system, length of pulse is important and, when transmission is weak or made under noisy conditions, the original signal may become so distorted as to be useless. With P.P.M. it is only necessary to sharpen the leading edge of each pulse by a suitable circuit and the i.c. can reconstruct the original pulse train for itself.

Another way in which this system overcomes interference is that the i.c. waits until it has received two *identical* pulse groups before it acts. If there is an error (perhaps caused by interference) in the first pulse train to arrive, this does not lead to an incorrect operation being performed. The first pulse train is compared with the second to arrive and, *if* they are identical, action is taken. If not, the i.c. rejects the first, awaits the third train and compares this with the second, and so on. Under bad reception conditions this could cause a slight delay in response but this is preferable to wrong functions being activated and is a warning that perhaps the control operation should be terminated.

The i.c. is connected as in Figure 14.1. The frequency of the internal clock is determined by the values of C1, VR1 and R2, according to the equation:

$$f = \frac{1}{0.15C_1\,(R_1 + R_2)}$$

where VR1 has value R_1 and R2 has value R_2. R_1 must lie

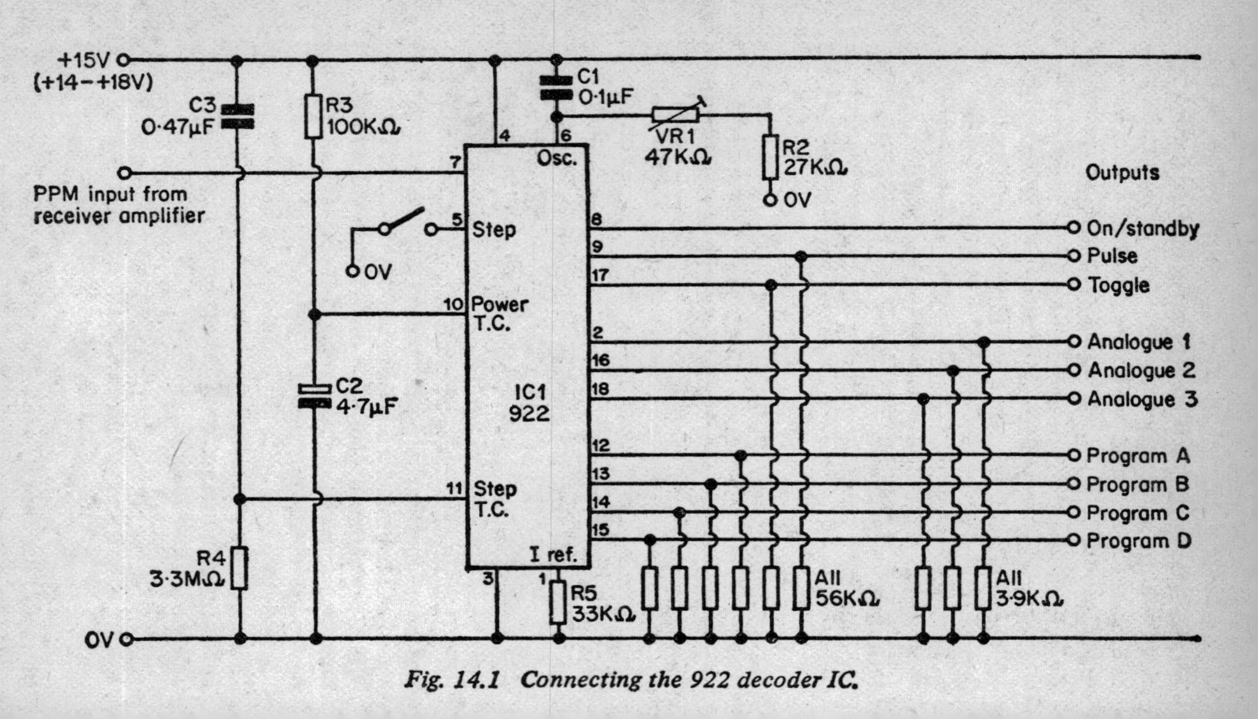

Fig. 14.1 Connecting the 922 decoder IC.

between 25kΩ and 200kΩ. It is to be adjusted in value so that $f = 40/t_0$, t_0 being the time for a '0' interval, as explained in Chapter 13. C2 and R3 determine the time taken for the 'initial' conditions of outputs (see later) to be set up following the switching on of the i.c. The values given make this period 2 second. C3 and R4 determine the *step time constant.* This determines how rapidly the program outputs are stepped through their sequence of programs when the 'program step' (+ or –) command is received. The program outputs can also be made to step by a switching action at the receiver. If pin 5 is grounded, the program outputs step through their sequence, cycling continuously at the rate determined by the step time constant. This function could be remotely controlled by using one of the analogue outputs.

The analogue outputs each produce a current that is a given fraction of the reference current, I_{ref}. The value of I_{ref} is determined by resistor R5. With the value shown, I_{ref} is between 250 and 450µA. At switch-on (initial conditions) the current from analogue outputs is 12/8 of I_{ref}, that is to say somewhere in the range 375 to 675µA, and with the loads represented by the 3.9kΩ resistors the voltages at these outputs are in the range 1.5V to 2.6V. As the 'analogue +' command is transmitted, output currents increase in steps of 1/8 of I_{ref} up to a maximum value of 31/8 of I_{ref}. As the 'analogue –' command is transmitted, the output currents decrease by steps of 1/8 of I_{ref} to zero.

The program outputs are listed in the table below. As can be seen, they are simply the inverse of the corresponding program number, in binary form:

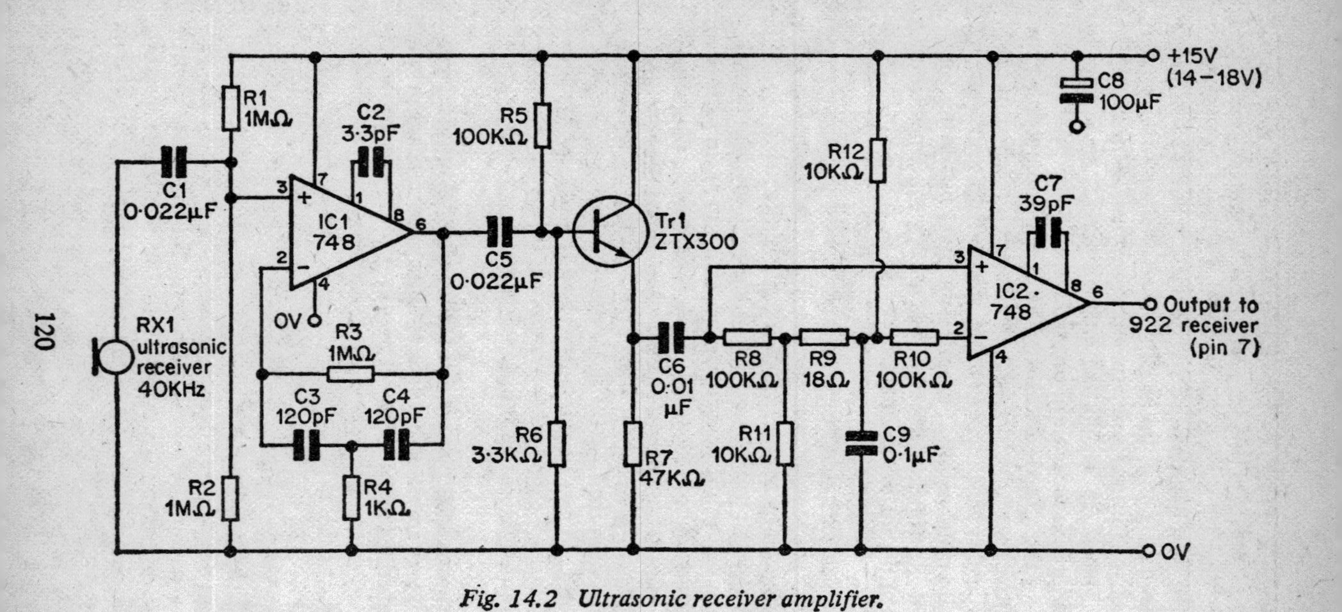

Fig. 14.2 Ultrasonic receiver amplifier.

Program number	Outputs			
	A	*B*	*C*	*D*
0	H	H	H	H
1	H	H	H	L
2	H	H	L	H
3	H	H	L	L
4	H	L	H	H
5	H	L	H	L
6	H	L	L	H
7	H	L	L	L
8	L	H	H	H
9	L	H	H	L

L = low H = high

Feeding signals to the decoder i.c.
The way this is done depends on the transmission method employed. It is possible to use line transmission of commands, in which case a low-value capacitor should be used between the terminal of the line and the input to the decoder. A 0.001μF capacitor is suitable. In this application, there is no carrier frequency, so the coder is without C1, and a fixed resistor 2.2kΩ is used in place of R1 (see Figure 13.2). For infra-red and visible-light transmission a high-gain amplifier such as that shown in Figure 7.6 can be used.

With ultra-sonic transmission the circuit of Figure 14.2 provides for demodulation of the signal to make it suitable for the receiver i.c. The demodulated signal from the interface of the radio receiver circuit (Figure 16.2) is also suitable for feeding to this i.c.

Using the i.c. outputs
(1) Program outputs:
These outputs are intended primarily for switching channels on a remote control television set. They have the same characteristics as the outputs 2 to 7 of the remote control radio circuit of Chapter 11, in that only 1 channel or program

is to be on at any given time and that selecting a new program automatically cancels the previous selection. If you have a model that is to perform any one of 10 functions but only one function at a time, the logic of the i.c. provides ten different 4-digit binary numbers that simply need decoding. If the outputs are used direct, we have a run from binary 6 to binary 15 (see table) and the obvious choice for a decoder is the CMOS 4514 or 4515 i.c. These are binary to 1-of-16 decoders, the 4514 giving a high output on the selected output line and the 4515 giving a low output on the selected lines, all other lines having the reverse state. A circuit for decoding appears in Figure 14.3. The outputs on lines 6 to 15 may then be used to drive lamps, motors, relays and other devices as described in Chapter 11. Remember to leave one code free as an 'all-off' code should you want to have the state in which no functions are in operation. Another way of doing this is to connect 'standby' output to the 'inhibit' input of the 4514 or 4515. This is initially high, so the initial program output (1111) is inhibited. As soon as a program command is received 'standby' output goes low, allowing the decoder to

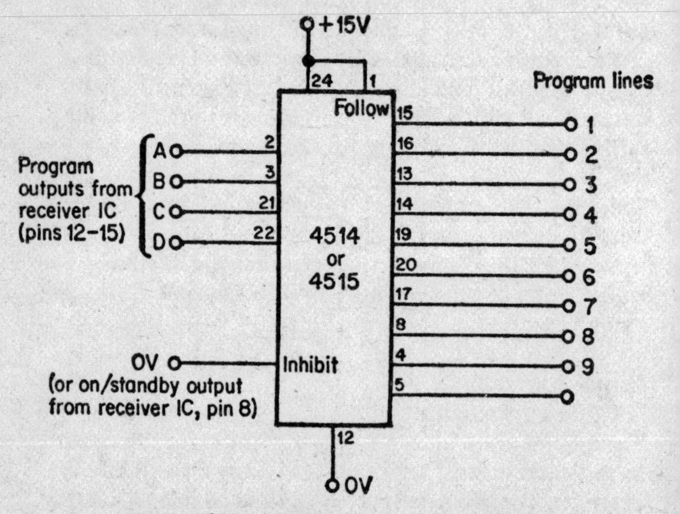

Fig. 14.3 Decoding the program outputs.

produce the appropriate output. On pressing 'standby' the 'standby' output goes high again, thus inhibiting the output from the decoder and causing all functions to cease.

If 4 or fewer functions are to be controlled by program outputs, a decoder is unnecessary. Each function is controlled by one of the program outputs A, B, C and D. Using invert gates if necessary, the functions are brought into action when the corresponding line goes low. Thus we can command program 1 to make line D go low, program 2 to make line C go low; program 4 to make line B go low; and program 8 to make line A go low. These lines could instead be connected to two flip-flops giving toggle action on two functions (Figure 14.4). Program 8 provides on–off control for one function; program 4 provides on–off for the other function. These are but a few examples of the ways the

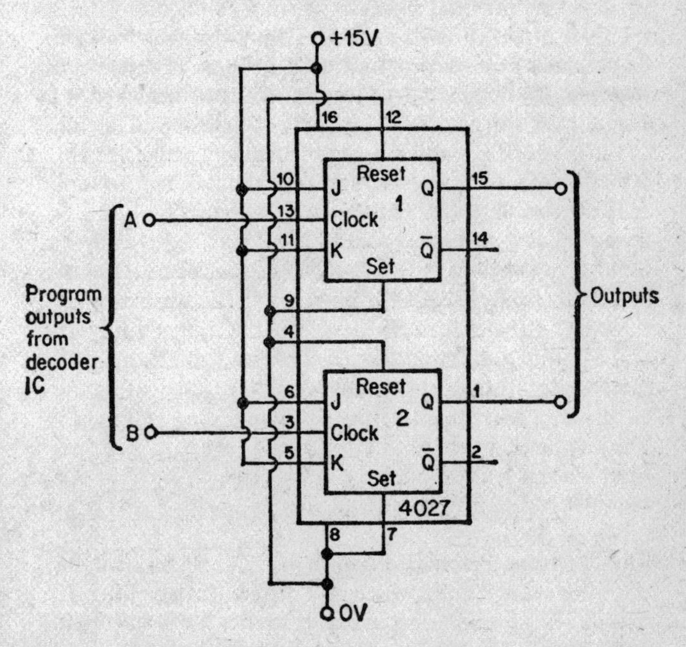

Fig. 14.4 Using a 4027 IC to produce toggle action on program outputs. Inverse action is obtained by connections to pins 2 and 14.

program outputs can be used. In any given circumstance it is generally a simple matter to devise the required logic, based on the principles discussed here and elsewhere in the book.

(2) Analogue outputs:

The most obvious uses for these outputs are the control of analogue functions such as the speed of a motor, the volume of sound from a radio, the brightness of a lamp, and so on. With the value of R5 given in Figure 14.1, the current from each analogue output can be controlled in the range zero to approximately 1.3mA. If the sink resistors have the value of 3.9kΩ as shown, this gives a potential at each output that can be varied from zero to 5V. With resistors of greater value, higher potentials can be obtained, but there is loss of linearity.

Analogue outputs will normally be required to control devices requiring higher currents than the 1.3mA maximum that can be obtained from each output, so some type of interface is required. The simplest type consists of transistors wired as in Figure 11.6. An analogue output may be connected directly to the base of TR1, and IC1 and the resistors connected to it may be omitted. TR1 requires only about 0.2mA to turn it fully on, so will be turned fully on by a current of 4/32 I_{ref}, giving only a 4-step speed control. This may be advantageous in some applications but control will be too coarse for most applications. By increasing the value of R5, we can reduce I_{ref} to a suitable value. A value of 180kΩ is a good starting-point in this connection though, since i.c.s transistors and motors vary in their characteristics, it is best to set up the circuit on a breadboard before finally deciding on values. The servo-mechanism circuit of Figure 11.9 can also be driven from the analogue outputs, the output being fed to R9. Again, the value of I_{ref} may need reduction.

Another output interface is shown in Figure 14.5. Here the operational amplifier is wired as a differential amplifier, its output being proportional to the difference between currents flowing to its two inputs. The current to the inverting input is controlled by adjusting VR1, which can be set so as to give

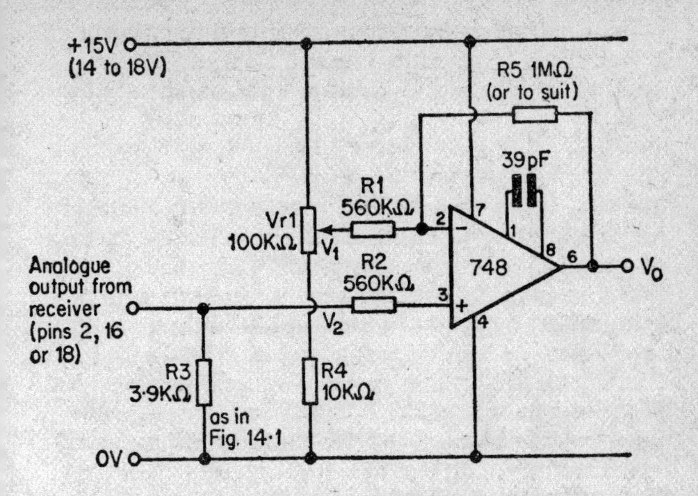

Fig. 14.5 An interface for the analogue outputs.

zero output at any required level. In particular, the output can be set to be zero when the current from the analogue output is 12/8 of I_{ref}. This is the value always taken at switch-on and on receipt of the 'normalise' command. In this way we can reduce motor speeds instantly to zero by sending the 'normalise' instruction, rather than by reducing speed over a period of a few seconds by using the 'Analogue –' command. The maximum rating of most operational amplifiers is 500mW which is less than most low-voltage motors work at when running under load. A rating between 1W and 6W is common.

It is therefore necessary to drive the motor by using a power transistor. Figure 14.6 shows a straightforward way of doing this; this method can also be used for brightness control of lamps and other applications. Figure 14.7 shows a circuit for motor control that is particularly useful when the motor is subjected to varying load during operation. If load on the motor increases, current through the motor increases, causing a rise in potential at the junction between the motor and the emitter of the transistor. This rising potential is fed to the inverting input of the amplifier, causing its output to fall, switching the transistor slightly off and reducing the flow of

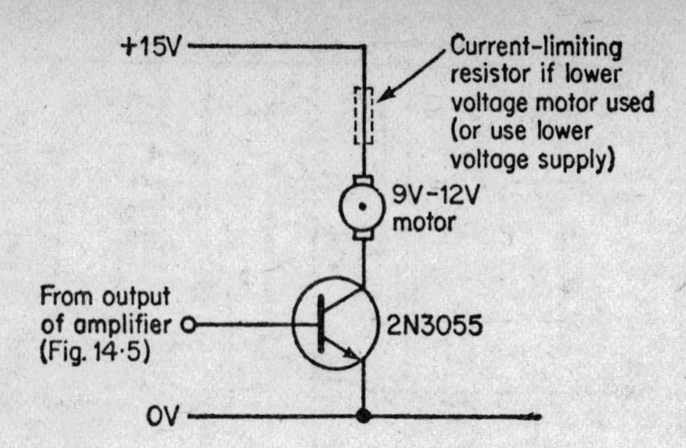

Fig. 14.6 Controlling a low-voltage DC motor.

current to the motor. Thus, this circuit maintains a steady current to the motor, keeping it at steady speed under varying load. It also allows the motor to run steadily at low speed, without jerking, and gives greater realism to models powered in this way.

In audio applications the analogue outputs can be used for the control of volume. With three analogue outputs it would be possible to control two channels of a stereophonic system independently, and use the third output for balance control. A device for controlling volume is the MC334OP electronic

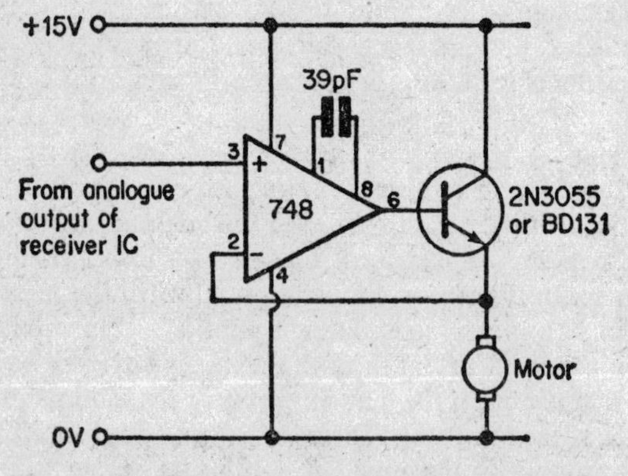

Fig. 14.7 Stabilized speed control for a DC motor.

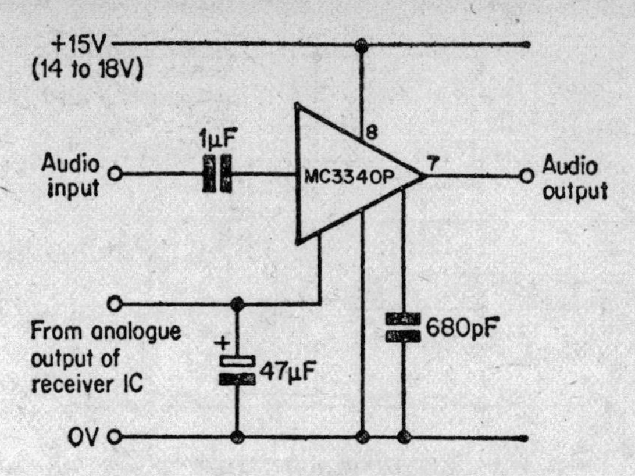

Fig. 14.8 Controlling volume levels in audio remote control applications.

attenuator (Figure 14.8). By varying the current supplied to its control pin (pin 2), the audio signal may be attenuated in the range 0 to 90dB.

(3) On/Standby output:
This is initially high, goes low whenever a program change is commanded and may be made high by 'standby' command. It then remains high until another program change is commanded. This could be used to control a muting operation in an audio application. The sound would be muted by the 'standby' command, and restored whenever a new program was selected. In model control it could be used to switch off power to several sections simultaneously pending the activation of one or more sections by program outputs. In other words – a true 'standby' control, putting the model temporarily in an inactive condition while awaiting fresh instructions. The logic necessary for making use of this output could be as simple as a CMOS buffer gate, controlling whatever power transistors are needed to effect the required actions.

(4) Pulse output:
This produces a low pulse of length approximately equal to $1/f_{osc}$ whenever any command signal is received. One

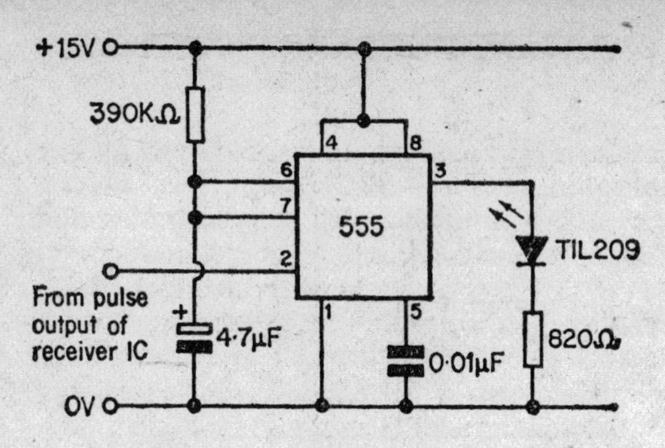

Fig. 14.9 Using pulse output to give indication that a signal has been received.

application of this is some visual indication that the signal has been received. This could take the form of an LED which flashes to indicate that event. Figure 14.9 shows how a 555 timer i.c. can be triggered to flash an LED once on receipt of a pulse from Pulse Output. The length of flash depends on the values of C1 and R1. With the values given the period is approximately 2 seconds.

(5) Toggle output:

This output is low at switch-on. If 'Analogue 2' output is reduced to zero, 'Toggle' output goes high. It remains high for as long as 'Analogue 2' output is zero; if 'Analogue 2' output is increased, either by using the 'Analogue 2+' command or by the 'Normalise' command, 'Toggle' output returns to low. When 'Analogue 2' output is greater than zero, 'Toggle' output can be made high by sending 'Toggle' command. It can then be made low again by sending any program command, any analogue command, 'Standby' or 'Normalise'. The change from high to low occurs immediately *after* the command has been effected.

15 RADIO CONTROL TRANSMITTER

Although infra-red and ultra-sound systems are ideal for short-range control because of their relative simplicity and the fact that no transmitting licence is needed, radio control is virtually essential for control of model aeroplanes and preferable for boats and yachts. Ready-built transmitter units are available commercially but for those who wish to build their own, Figure 15.1 gives a circuit that is simple to construct and highly reliable in operation. It operates on a radio frequency in the 27MHz band and is 100% amplitude modulated. The modulation may be effected either from a TTL output, from a CMOS output or from the 490 coder i.c. described in Chapter 13. Thus, any of the systems described in this book can be used with this transmitter.

Transmission within the 27MHz band is restricted to a number of defined frequencies spread more or less evently across the band. Frequency-controlling crystals are sold in matched pairs for the six main channels, each of which is designed by a colour: brown, red, orange, yellow, green and blue. If you are likely to be operating your transmitter at the same time and in the same locality as other transmitters it is essential that no two transmitters are operating on the same frequency. If the transmitter and receiver are fitted with sockets, crystal pairs may be changed at will, so as to operate on a vacant frequency. Note that the frequency of the receiver crystal is 455kHz lower than that of the transmitter crystal, since this is the intermediate frequency used in the receiver circuit. It is essential to use the correct crystal when assembling the circuit.

The circuit may be assembled on stripboard or on a printed circuit board. A special circuit-board is available for this transmitter (see p. 164). Three coils are needed for the transmitter, and these are constructed as follows:

L1 consists of 10 turns of 30 s.w.g. enamelled copper wire close wound on a coil-former of diameter 7mm. The coil is centre-tapped; this is more easily done if the length of wire

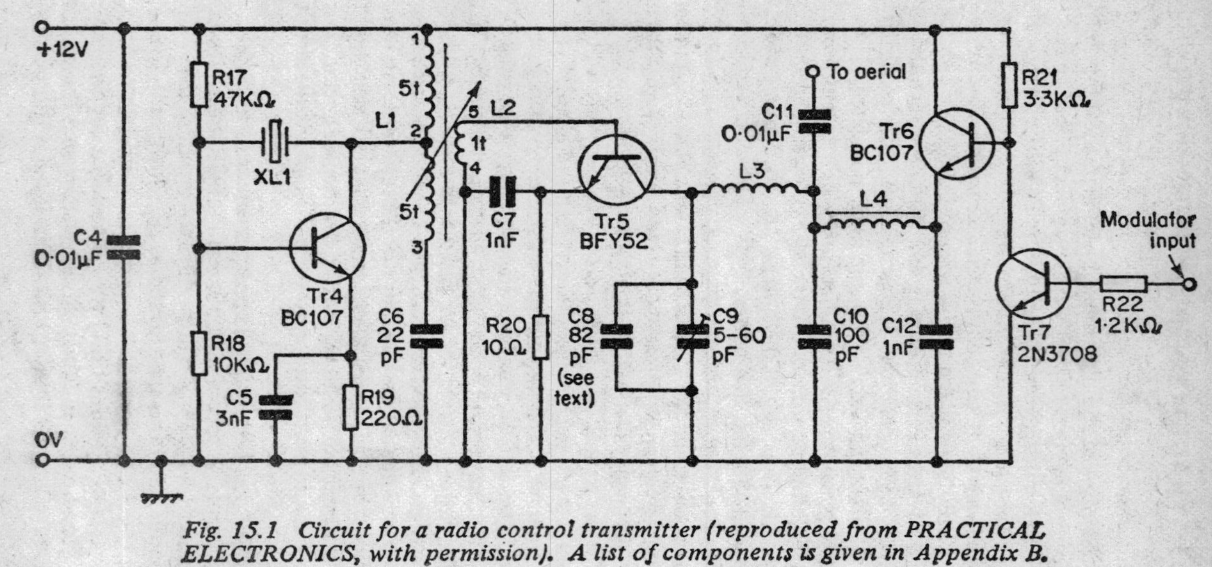

Fig. 15.1 Circuit for a radio control transmitter (reproduced from PRACTICAL ELECTRONICS, with permission). A list of components is given in Appendix B.

intended for the coil is scraped for a few millimetres about half-way along its length. A short piece of wire of the same kind is then looped around the bare portion and soldered as neatly as possible. The coil is then wound 5 turns in each direction from this point. Note that the direction in which the coil spirals around the former must be constant for both halves of the coil. The coil former should have a threaded iron dust core to fit.

L2 is a single turn of wire of the same gauge on top of L1 at the positive end.

L3 is an open coil of 9 turns of 14 s.w.g. wire. The inside diameter of the coil is 16mm and the windings should extend for a distance of 38mm. To wind this coil, find a smooth cylindrical object that is 16mm diameter (some makes of thick felt-tip pens are this size) and wind the coil around it, trying to make it about 38mm long. Then remove the cylinder and secure the coil close to the circuit-board, soldering its ends in holes previously drilled exactly 38mm apart.

Power supply

The transmitter circuit requires a power supply at 12V, and takes about 40–60mA. Most users require that the transmitter be portable, so a battery of 8 dry cells, 1.5V each is required. If coding is done by TTL this requires, ideally, a 5V supply. Since TTL can also work on 6V it is simpler to centre-tap the 12V battery and draw the TTL supply from this. Similarly, the battery may be tapped at 9V to supply CMOS or the 430 coding circuits.

Case and aerial

The case needs to be large enough to contain the battery, coding circuit and transmitter circuit. It should have a panel large enough to accommodate control keys, joy-sticks (if used) an on–off switch and any indicating lights that are required. The case should be substantial enough to withstand usage outdoors and to survive being dropped. A telescopic aerial is required. It should be 1 metre long and be mounted so as to be vertical when the case is being held in its normal operating position.

Setting up the transmitter circuit

Turn the core of coils L1 and L2 to bring it to the top end of the former. Set C9 – the trimmer capacitor, to its maximum. Connect a voltmeter (about 3V full scale deflection) across R20. Then switch on the 12V supply. Adjust the position of the core until the voltmeter reading is a maximum. Then adjust C9 until the voltmeter reading is maximised. It may be found that the maximum reading is beyond the range of adjustment of C9. In this event exchange C8 for a capacitor of higher or lower value. Final tuning of the transmitter can be carried out with the circuit mounted inside its case, when the receiver circuit is ready for testing.

Licence

A licence is required to operate any radio control system. The licence, which lasts for 5 years, may be obtained on application to: The Home Office, Radio Regulatory Department, Waterloo Bridge Road, London SE1 8UA. First write to this address requesting an application form, then send the completed form together with the fee. The fee at present is £2.80.

16 RADIO CONTROL RECEIVER

The circuit is shown in Figure 16.1. Construction is simplified by employing the TBA651 tuner and intermediate-frequency amplifier i.c. as the heart of the circuit. Coils L1 and L2 are made as follows:

L1 consists of 13 turns of 28 s.w.g. enamelled copper wire, wound on a former diameter 7mm, as used for the transmitter (Figure 16.2).

L2 consists of 4 turns of the same wire wound on the same former a short distance from L1.

Layout of this circuit should be as compact as possible, for usually it is to be used in a model vehicle where space is at a premium. A special printed circuit board is available ready-made for this receiver (see p. 164).

Power supply

The circuit takes only 10mA and operates on a 9V supply, so the circuit itself requires only a small battery such as a PP3. However, the decoder circuit used may well require more than this, and the various motors and other controlled devices will probably require considerably more. If different voltages are required for different parts of the system, use a battery pack to supply the highest voltage and tap this to obtain the lower voltages. It is well worth while to decouple the power supplies to each part of the circuit by connecting a capacitor (0.1μF or more) across the positive and negative lines where they enter each circuit board.

Aerial

The aerial should be as long as convenient, up to 1 metre long if possible. Often this is *not* possible since the model can not support it. The aerial may be made from stout copper wire and is usually placed so as to resemble a radio antenna roughly in scale with the model.

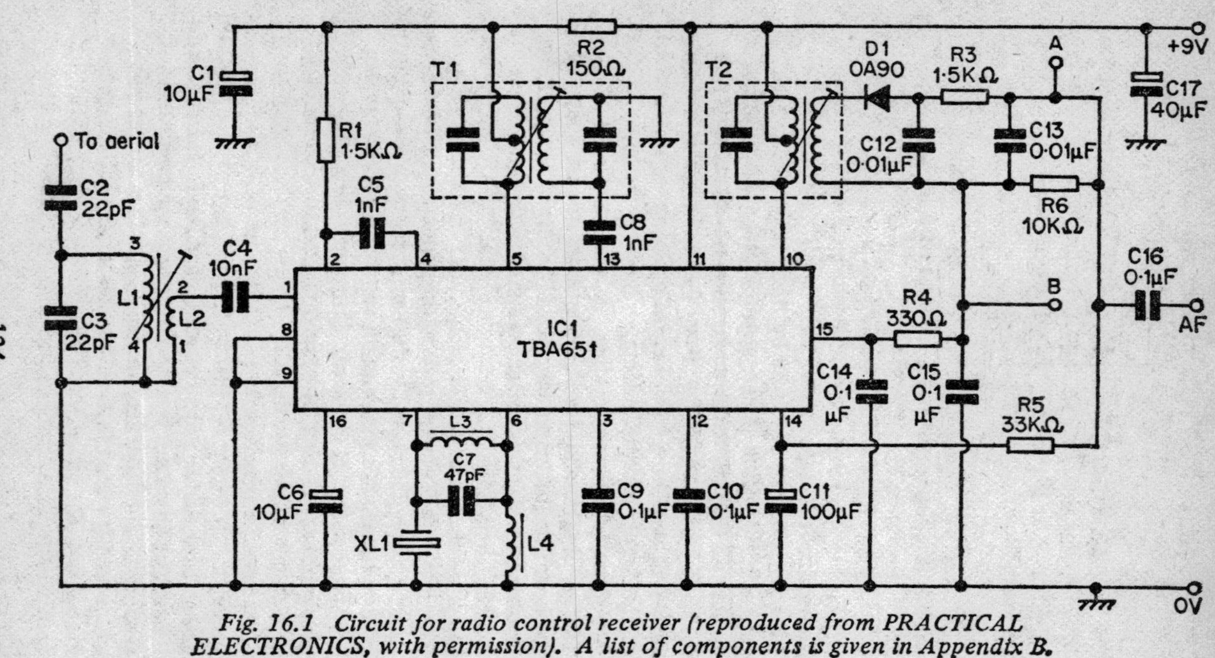

Fig. 16.1 Circuit for radio control receiver (reproduced from PRACTICAL ELECTRONICS, with permission). A list of components is given in Appendix B.

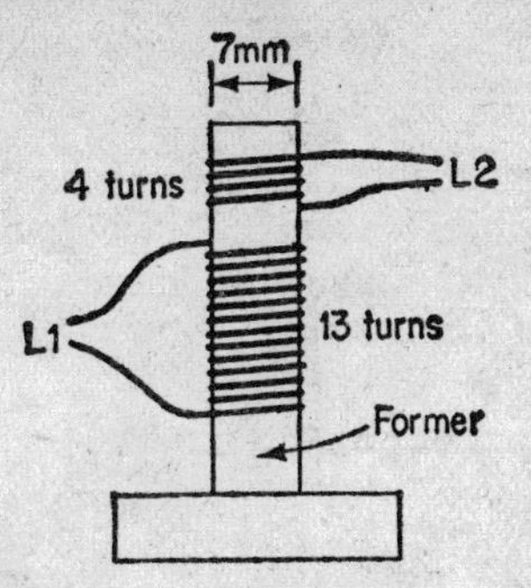

Fig. 16.2 Winding coils L1 and L2.

Testing the receiver

Check that the current taken by the receiver circuit when connected to a 9V supply is approximately 10mA. Next connect a low-range voltmeter across points A and B, Figure 16.1. Place the transmitter a few metres from the receiver with its aerial disconnected and switch it on. Adjust the core of T2 to give maximum reading; then adjust the core of T1 to give a further maximum reading. Repeat the adjustments to T2 and T1 alternately, until the greatest attainable value is reached. Next adjust the core of L1/L2 to obtain the greatest possible further increase in voltage on the meter.

Decoder interface

The circuit of Figure 16.3 responds to positive-going pulses by producing a high output. At the negative-going edge its output falls low again. This interface is suitable for all the coding systems described in this book. It may be operated on a 5V supply if the decoder is based on TTL i.c.s. If CMOS i.c.s are used, the interface and decoder may be operated from the same supply as the receiver (9V). For use with the 922 decoder i.c. (Chapter 14) the operating voltage may be 15V. If pulses are relatively long and the true pulse form is required for proper decoding, as with the multiple-pulse system (Chapters 8 and 9), it may be necessary to increase the value of the coupling capacitor of the receiver circuit, C16, Figure 16.1. This may be increased to 47μF by wiring a 47μF capacitor in parallel with C16 so as to produce long pulses that are detectable on a voltmeter connected to the output of the interface circuit.

The interface circuit looks rather complex and space-consuming, but with a certain amount of planning plus skill with the soldering-iron, it can easily be accommodated on a piece of stripboard only 1.4cm x 3.8cm.

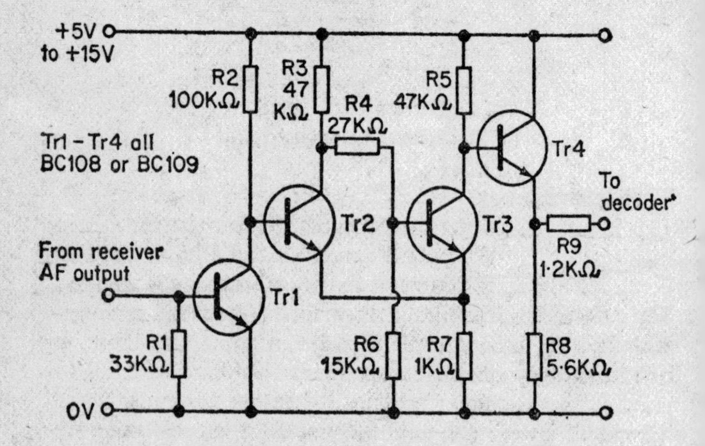

Fig. 16.3 Interface circuit.

Radio-controlled model boat

As an example of the planning and designing of a radio-controlled model, we will consider a system developed for a model boat, powered by an electric motor. The boat itself was commercially made from plastic; originally it had been controlled remotely by a power-line from a control box, held by the operator on the bank. Steering was effected by a second electric motor, geared to the rudder.

It was decided to use the coder and decoder i.c.s described in Chapters 13 and 14. The following functions were listed and allocated as shown:

Function	*Action*	*I.C. code*	
Propulsion motor	Off	Normalise (with analogue 1 at 12/8 I_{ref} motor speed is to be zero; ensures speed is zero at switch-on also).	
	Forward	Program 0	
	Reverse	Program 1	
	Increase speed	Analogue 1+	
	Decrease speed	Analogue 1–	
Steering motor	Left (port)	Program 2	To increase or decrease the amount of turn at each transmission of the signal
	Right (starboard)	Program 3	
Navigation and other lights	On/Off	Program 4 (toggle action at each signal)	
Hooter (buzzer)	Single blast	Program 5	

Only six programs are being used, so it is not necessary to decode output D (see table, p. 121). Three-line decoding can be done with the 4028 i.c., provided that the outputs from the receiver i.c. are inverted before decoding. This avoids the expense of using the 24-pin 4514 or 4515 i.c.s.

The control panel requires 9 key-switches, which may be conveniently arranged as in Figure 16.4. The transmitter requires a 12V supply, consisting of eight 1.5V cells which may be contained in two ready-made 6V battery-holders. A thin sheet of metal may be placed between terminals of adjacent batteries to tap this supply at 9V, as required for the transmitter i.c. The keys (Figure 16.5) are indicated by circles, and call the code groups listed in the table above.

The receiver decoder and control circuits are given in Figure 16.6. The propulsion motor is supplied with current by way

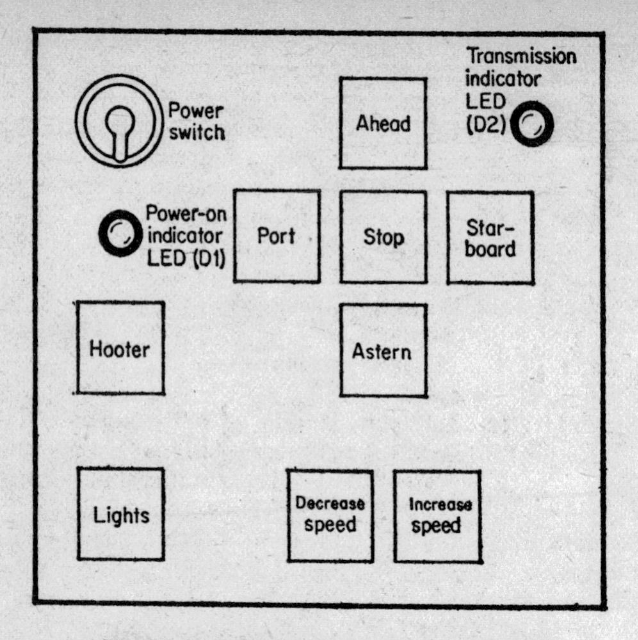

Fig. 16.4 Control panel for remote control transmitter for radio controlled motor boat.

of a 2-pole changeover relay. When energised the direction of current is such as to make the motor propel the boat astern. When the relay is de-energised the boat moves forward. This arrangement means that the relay is energised only infrequently and for short periods, thus saving current. The current supplied to the propulsion motor is under the control of analogue 1 output. By suitable choice of values for R1 and VR1, and by correct adjustment of VR1, the supply of current may be set so as to be just insufficient to drive the motor when Analogue 1 output is at 12/8 I_{ref}.

The steering motor is supplied with a short burst of current whenever Program 2 *or* Program 3 is received. When Program 3 is received (but *not* when Program 2 is received) a similar pulse energises Relay 2. If energised, the relay switches current so as to give a right-ward turn to the boat.

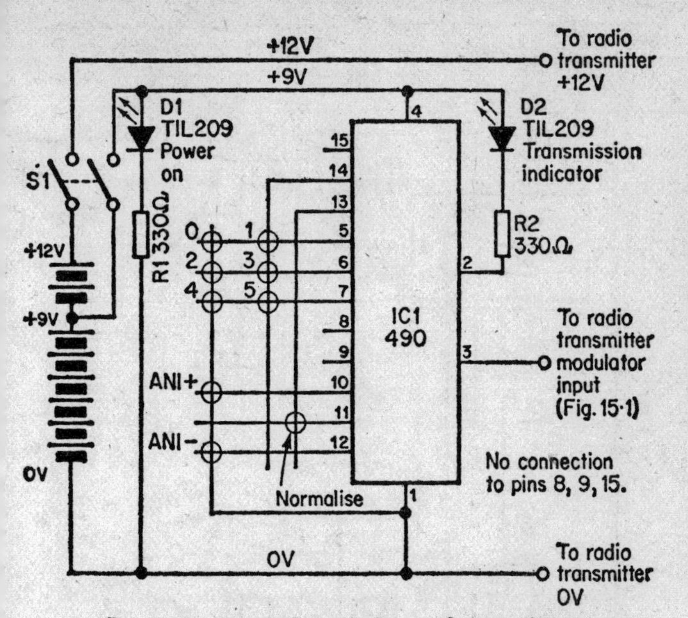

Fig. 16.5 Circuit for radio control transmitter.

If not energised, a left-ward turn is produced. This arrangement requires the reversing relay to be energised only when there is a *change* of rudder position in a right-ward direction, which means that very little current is consumed for the steering function.

The lights on board consist of a red LED for port navigation light, a green LED for starboard navigation light, and one or more filament lamps (and further LEDs, if required) as cabin lights, spot-lights and so on. These are controlled by way of a J-K flip-flop (IC6) that changes state each time it receives a positive-going pulse edge. The lights are switched on or off alternately each time Program 4 command is received.

The buzzer, which is a semi-conductor device that gives an extremely loud noise yet requires relatively small current (about 10mA) is obtainable from Maplin Electronics Ltd. (p. 164). Note that the red wire and black wire indicate to

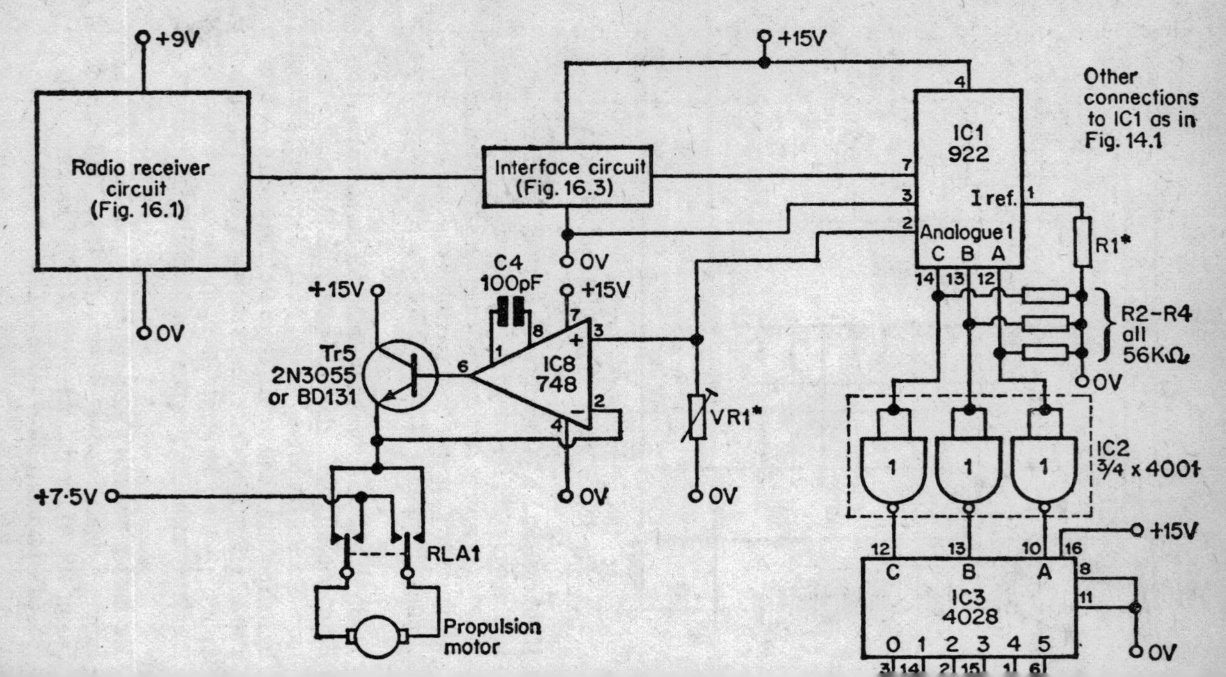

+9V
Radio receiver circuit (Fig. 16.1)
0V
+15V
Interface circuit (Fig. 16.3)
0V
IC1 922
I ref.
Analogue 1
C B A
Other connections to IC1 as in Fig. 14.1
R1*
R2–R4 all 56KΩ
0V
C4 100pF
+15V
+15V
Tr5 2N3055 or BD131
IC8 748
VR1*
0V
0V
+7·5V
RLA1
Propulsion motor
IC2 3/4 x 4001
+15V
IC3 4028
0V

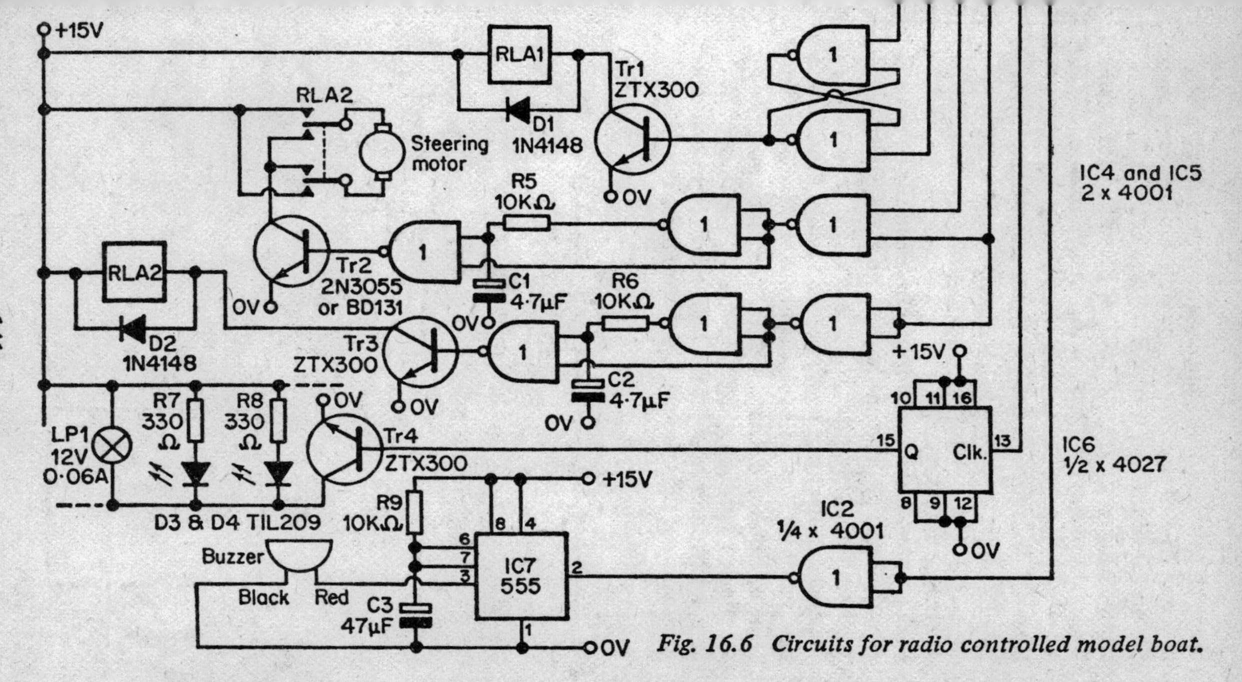

Fig. 16.6 Circuits for radio controlled model boat.

the polarity of this device and *must* be connected as shown in Figure 16.6. The power supply to the buzzer comes from the output of the 555 timer i.c. This is wired as a monostable multibrator. It produces a single pulse, lasting about 2 seconds, whenever its input is triggered by a low-going pulse. The length of pulse can be changed by altering the values of R9 or C3. It is possible to run several buzzers connected in parallel to the output of IC7 if a louder noise is required.

Alternative decoding

Such a system as shown in Figure 16.6 could also be controlled by the multiple-pulse method described in Chapters 8 and 9. The main disadvantage is the increased number of i.c.s required and the extra power needed to drive them. For a land vehicle or large boat this would not be greatly disadvantageous. Control of the speed of the propulsion motor could be effected by using a digital-to-analogue converter. A completely different approach, with a marked reduction in the number of i.c.s required, is discussed in the next chapter.

17 REMOTE CONTROL MICROPROCESSOR

A glance at Figures 8.4, 9.1, 11.11 or 16.4 shows that, even for a relatively limited number of control functions, a complex circuit is required. The number of i.c.s increases by twos and threes for every additional function. Not only does this mean that more space is required *and* that more current is required (with consequent increase in weight, and the expense of renewing spent batteries) but the circuits themselves become too unwieldy. With increasing complexity there is greater risk of spurious triggering through almost unpredictable and undetectable pulses appearing inexplicably. There is the problem of 'races' when gates may not operate in exactly the right order to prevent unwanted states from occurring, and when transitions between one state and another may cause illegal transition states to appear, so preventing the system from functioning in the logical way its design requires. It is true that by careful design many of these problems can be eradicated, but there is a limit to the extent to which design can reduce the number of i.c.s needed to control a given set of functions. Indeed, it generally happens that the design features required to eliminate races and transition-state problems actually lead to an *increase* in the number of i.c.s!

At this stage we may adopt an entirely different strategy, with the help of the microprocessor. Instead of increasing the number of i.c.s (increased hardware), we can base the control system on a microprocessor and the few peripheral i.c.s it needs, and then program the microprocessor to carry out all the logical operations formerly carried out by the logic i.c.s. We substitute software for hardware. A microprocessor is a multi-purpose logic i.c. and when correctly programmed it can perform all the operations required for a remote control coder or decoder. It can also provide a carrier frequency if required. It can perform the other logical operations such as latching, pulse generation, and toggle-action, for switching certain functions. Apart from a keyboard at the transmitter and power transistors at the receiver nothing more than the microprocessor i.c. and its few associated i.c.s are required.

By using a method of pulsed current supply (through power transistors) the microprocessor can also control analogue functions such as motor speed.

There is not space here to go into full details of such systems, but a few examples show the way in which this type of control may be effected. The details depend on the type of microprocessor being used, but the program flow charts should help the experimenter to work out a program for almost any type.

Figure 17.1 shows the program flow chart for a program that carries out the same functions as the circuit of Figure 8.4. If you have even a simple microprocessor system such as the *Science of Cambridge MK-14* or the *Acorn*, it can be programmed to do exactly the same task as the circuit of Figure 8.4. The operator presses one of the keys of the microprocessor's keyboard and one of the output ports of the system produces a pulse train that is coded according to the system described in Chapter 8. The output port is compatible with TTL or can feed base current to a switching transistor, so that the system can readily be interfaced with an infra-red LED, a filament lamp, the radio transmitter of Chapter 15, or the ultra-sound transmitter of Chapter 3. In the latter two cases, the output can be fed directly to the circuit and no interface is required.

The operation of the program is simple. In the diagram, x represents the state of an output port that has been chosen for use with the remote control system. In the program for the MK-14 (Appendix A, p. 151) the Flag 0 output is chosen; this output comes directly from the microprocessor i.c. itself, needing no input/output device. This is an advantage if one is contemplating building a special transmitter system (using the SC/MP microprocessor i.c.) rather than using the full MK-14 system. With other microprocessor i.c.s a special interface i.c. may be required; such i.c.s are usually manufactured to suit the requirements of a particular microprocessor i.c. The program is entered from a chosen START address and the first action is to make the

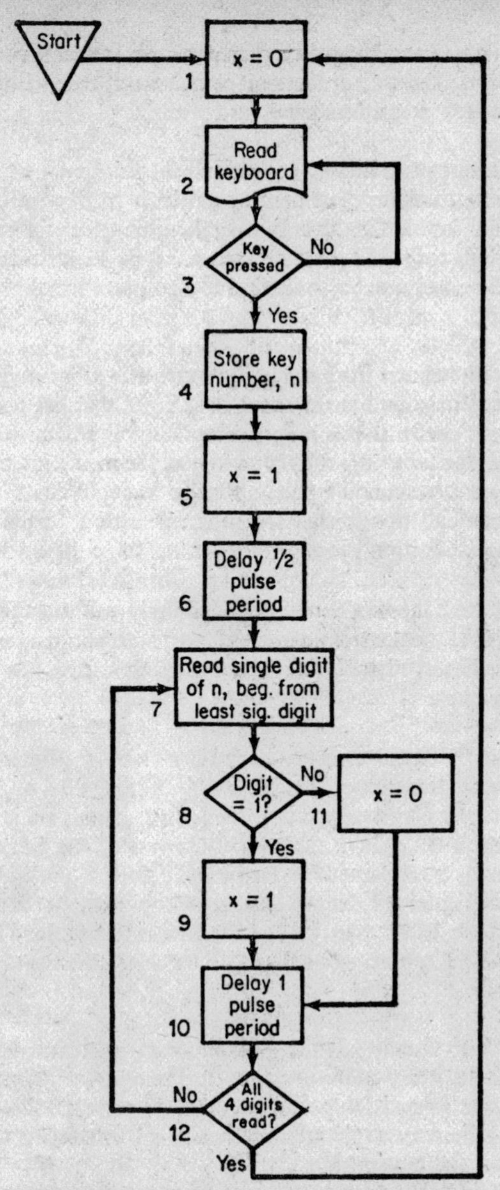

Fig. 17.1 Program flow chart for multiple-pulse remote control system – transmitter.

output low ($x = 1$, in box 1). The program then instructs the microprocessor to read the keyboard, until it detects that a key has been pressed. The number of the key is then stored in memory (box 4). Such procedures are routine and the way they are programmed depends on the instruction set of the microprocessor. The next operation is to transmit the 'start' pulse which is done by making output high ($x = 1$) and waiting for half a pulse period. Next, the least significant digit of the key number, n, is read to find out if it is 0 or 1. If it is 1, x is made high, but if it is 0, x is made low (boxes 8–11). A delay of 1 pulse period then follows, so producing a high or low pulse from the transmitter. The program needs provision for counting the number of digits that has been read and coded. If this number does not yet amount to 4 (box 12) the program loops back to read the next digit and sends the corresponding pulse. Finally when all digits have been read and transmitted the program returns to the beginning, awaiting the next instruction to be keyed in.

Once started, the program loops indefinitely until the operator decides that the control operation is to be discontinued and presses a 'reset' button that causes the microprocessor to leave this program.

As far as the operator is concerned it makes no difference whether the transmitter uses the circuit of Figure 8.4 or the microprocessor system just described. Instructions are keyed and transmitted equally easily by either system. As far as the constructor is concerned it is easier and cheaper to build the circuit of Figure 8.4 than to build a microprocessor-controlled transmitter. What, then, are the advantages to be gained by using the microprocessor system? The chief advantages are as follows:

(1) If one already has the microprocessor system, using it for its many other purposes, including calculation and playing games, it is extremely simple to let it double as a remote-control transmitter, and the interfacing need cost virtually nothing.
(2) In using the microprocessor as described, we are far from

making maximum use of its capacity. With a very few extra stages to the program it can operate an 8-digit or a 16-digit transmission system, giving scope for an extremely wide range of functions and allowing simultaneous control of very many channels.

(3) The program can be easily modified in the light of experience or if requirements alter, simply by re-writing portions of the program. It is less easy to re-wire a system based on dozens of integrated circuits. Thus its use becomes justified or even made essential if an elaborate remote control system is to be constructed.

(4) When part of a *receiver* system on a model in which space is limited, and especially when weight must be kept as low as possible, the advantages of a microprocessor-based system are even more obvious.

Microprocessor receiver program

The 4-digit register circuit of Figure 9.1 requires eight i.c.s and a considerable number of timing capacitors and resistors. Its output must then be fed to decoding circuits, so adding to the number of i.c.s needed. To extend its capacity to eight or more digits would require even more i.c.s making it too big to put inside, say, a model aeroplane. If we use a microprocessor we need only the microprocessor i.c. plus a 'clock' crystal, one or two memory i.c.s (in which the program is stored) and possibly (depending on the microprocessor employed) an interface i.c. The memory i.c.s would need to have the program copied into them, using a programmer circuit. Such circuits are available for use with most microprocessor systems and are cheap. There is not space here to go into the details of how this is done, but a program is given (Figure 17.2 and Appendix A) to illustrate the essentials of a receiver-decoder program. When running this program, a microprocessor system may be remotely controlled, either by the transmitter-coder circuit of Figure 8.4 or by another microprocessor system running the program of Figure 17.1.

After having been started, the program waits until a high input appears at the input port ($x = 1$, boxes 1 and 2). It delays a period equal to 1 pulse and then reads x again. This coincides

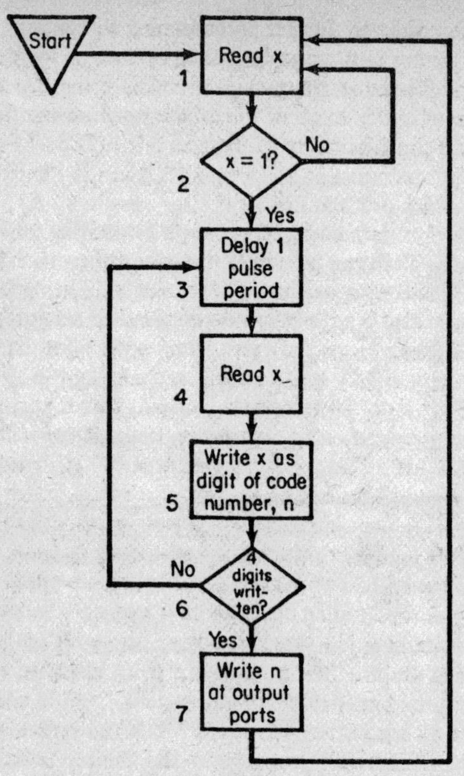

Fig. 17.2 Program flow chart for multiple-pulse remote control system – receiver.

with the centre of the first code pulse. The value of x (1 or 0) is then written into a memory which is to contain the 4-digit code group, n. The program loops back to box 3, waits for the next pulse, then writes its value into the next digit of the code group. This continues until all digits have been written, and the code group is complete. The group might then be written to four output ports of the system, causing them to become high or low. Alternatively as in the program in Appendix B the group is treated as one of 16 possible binary numbers and *one* of 16 corresponding outputs is made high. In short, the program can undertake the decoding function. By further programming we can arrange for certain of the outputs to be pulsed in various ways for varying periods of time; we can also arrange for toggle action at any given outputs. In other words, the program can fulfil the many different functions of the i.c.s IC2 to IC8 of Figure 16.4.

When we read of satellites and space probes having an 'on-board computer' we can be sure that a microprocessor is at the heart of it. It is programmed to receive and obey remote-control signals sent from stations on Earth. It is also programmed to exercise *local control* over certain operations that can not readily be remotely controlled. For example, it may be programmed to land the vehicle on a distant planet because delays in the propagation of radio waves make it impossible to control the action from Earth. This type of local control could also be programmed into a model control system. A vehicle could have sensors to detect objects in its path so that local control would over-ride remote control in such circumstances and the vehicle would not run into these objects. A remotely-controlled sailing boat can have a sensor that detects the angle of heeling; the program adjusts the position of a counterweight to bring the boat to the most effective angle with the vertical. In a model railway system it would only be necessary to key in the instruction 'Go from station A to station B' and the microprocessor would select the most suitable route, set the points and bring the train smoothly and safely to its destination. Such are the possibilities with microprocessors that the reader is left to explore. Perhaps it could be said that by leaving so much to the microprocessor we are taking a lot of fun out of the operation of remotely controlled models. This is true to a great extent, but if one's main interest lies in devising electronic control systems that are highly sophisticated, and more realistic in the actions they produce, this is certainly a new dimension in remote control that is as yet largely unexplored by the enthusiast.

APPENDIX A
MICROPROCESSOR PROGRAMS

Multiple-pulse coder

This program is the software equivalent of the circuit of Figure 8.4. It performs the same function, except that the digits are transmitted in the reverse order, l.s.d. first. It is written in machine code for the SC/MP microprocessor, as used in the *Science of Cambridge* MK-14 microprocessor system. To enter the program go to 0F20. The display shows all zeros. When an alphanumerical key is pressed the display blanks completely while the code group is being transmitted. When the zeros return the next command may be entered on the keyboard. The coded output is obtained from the Flag 0 output of the microprocessor i.c.

```
0F1E                  = digit counter, c
0F1F                  = key number, k
0F20   C40F           LDI    X'OF' ⎫
0F22   36             XPAH   P2    ⎬ Pointer 2 to
0F23   C400           LDI    X'00' ⎪ RAM (OF00)
0F25   32             XPAL   P2    ⎭
0F26   CA1E           ST     at c, clears counter
0F28   CA0C           ST     at ADH, clears 'address low'
                             display digits
0F2A   CA0E           ST     at ADL, clears 'address high'
                             display digits
0F2C   CA0D           ST     at WORD, clears 'word'
                             (data) display digits
0F2E   C401     A:    LDI    X'01' ⎫ Pointer 3 to display
0F30   37             XPAH   P3    ⎬ routine in monitor
0F31   C43F           LDI X'3F'    ⎪ (0140 minus 1)
0F33   33             XPAL   P3    ⎭
0F34   3F             XPPC   P3      go to display routine
                                     and wait until key
                                     pressed
```

```
0F35   922D          JMP to A       command key press-
                                    ed in error; program
                                    jumps back to here,
                                    so now jump to A to
                                    return to display
                                    routine
0F37   01            XAE     alphanumeric key pressed;
                             program jumps back to here
                             with number of key stored in
                             extension register; put this in
                             accumulator
0F38   CA1F          ST      number of key stored at k
0F3A   C401          LDI     '01'  } Make Flag 0 high
0F3C   07            CAS           }
0F3D   8FFF          DLY X'FF'      half a pulse delay
                                    (the start pulse)
0F3F   C21F    B:    LD      k
0F41   1E            RR      least sig. digit of k rotated to
                             m.s.d. position.
0F42   CA1F          ST      rotated k stored, ready for
                             next rotation
0F44   964A          JP      if l.s.d. is zero, to C
0F46   C401          LDI     X'01' } l.s.d. is '1', so make
0F48   07            CAS           } Flag 0 high
0F49   924D          JMP to D for delay
0F4B   C400    C:    LDI     X'00' } l.s.d is '0', so make
0F4D   07            CAS           } Flag 0 low
0F4E   8FFF    D:    DLY     X'FF' } full pulse delay
0F50   8FFF          DLY     X'FF' } (code digits)
0F52   AA1E          ILD     c, counting number of digits
                             coded
0F54   E404          XRI     X'04'   to test if c = 4
0F56   9E3E          JNZ to B, to obtain next l.s.d., if 4
                             digits not yet coded
0F58   C400          LDI     X'00' } coding completed,
0F5A   CA1E          ST      c     } clear c ready for
                                   } next transmission
0F5C   07            CAS     make Flag 0 low between
                             transmissions
0F5D   922D          JMP to A to await next command
```

The delay instructions at 0F3D, 0F4E and 0F50 each produce a delay of approximately 0.066s, with a crystal frequency of 4MHz. This means that the total time required for transmitting the command is about 9 times this length, a total of approximately 0.6s. Faster transmission may be achieved by reducing the delay periods. For example, if the three groups listed above are each altered to 8F2B, the transmission time is reduced to approximately 0.1s.

Multiple-pulse decoder

This program is the software equivalent of the circuit of Figure 9.1 *plus* a 4-line–to–16-line decoder such as the 4514 i.c. (Figure 14.3). It accepts any one of the 16 possible 4-digit codes produced by the multiple-pulse coder and produces a high output on the corresponding one of 16 output ports of the INS8154 input/output i.c. of the MK-14 microprocessor system, all other outputs being low. The basic program can easily be modified to produce any other pattern of outputs, as required. The command signal is fed to the SENSE A input of the microprocessor i.c.

```
0F1C                  =  digit    counter,  c
0F1D                  =  code     group,    n
0F1E                  =  output   code,  Port  A
0F1F                  =  output   code,  Port  B
0F20  C400   A:   LDI     X'00'  }
0F22  C8F9        ST      c      } Clear registers c, n,
0F24  C8F8        ST      n      } A,    and    B
0F26  C8F7        ST      A      }
0F28  C8F6        ST      B      }
0F2A  06     B:   CSA     to detect arrival of high pulse
                          (= start pulse)
0F2B  D410        ANI     X'10'   gives X'00' if input
                                  is low; X'10' if input
                                  is high
0F2D  98FB        JZ to B if input low, to await high
                          input pulse
0F2F  8FFF   C:   DLY     X'FF' } wait for whole pulse
0F31  8FFF        DLY     X'FF' } period
0F33  06          CSA     to read code pulse value
```

```
0F34  D410        ANI    X'10'   gives X'00' if pulse
                                 low; gives X'10' if
                                 pulse high
0F36  9C03        JNZ to D if pulse high
0F38  02          CCL make carry-link '0' because
                         pulse is low
0F39  9001        JMP to E
0F3B  03      D:  SCL    make carry-link '1' because
                         pulse is high
0F3C  C0E0    E:  LD n
0F3E  1F          RR     shift carry-link contents into
                         m.s.d. of n
0F3F  C8DD        ST rotated n, with new m.s.d., at n
0F41  A8DA        ILD c
0F43  E404        XRI    X'04'   to test if c = 4
0F45  9CE8        JNZ to C, to detect next pulse, if 4
                  pulses not yet registered
0F47  C0D5        LD n   Code group, n, stored in
0F49  1E          RR     high byte of 0F1D, is
0F4A  1E          RR     rotated until it is in the low
0F4B  1E          RR     byte
0F4C  1E          RR
0F4D  C8CF        ST n
0F4F  03          SCL    Inserts '1' as m.s.d. of A,
0F50  C0CD    F:  LD A   and shifts it one place to the
0F52  1F          RRL    right on each return to this
0F53  C8CA        ST A   step
0F55  C0C9        LD B   Receives l.s.d. from A and
0F57  1F          RRL    shifts it one place to the
0F58  C8C6        ST B   right on each return to this
                         step
0F5A  02          CCL    subsequent digits shifted into
                         A must be '0'
0F5B  B8C1        DLD n  counting n down to deter-
                         mine how many shifts must
                         be done
```

```
0F5D  94F1    JP to F  if shifting not yet complete;
                       when shifting complete,
                       either A or B contain a single
                       '1' in a position that depends
                       on the original code group, n
0F5F  C40A    LDI   X'0A' ⎤ Pointer P1 to
0F61  35      XPAH  P1    ⎬ input/output device
0F62  C400    LDI   X'00' ⎥ (0A00)
0F64  31      XPAL  P1    ⎦
0F65  C4FF    LDI   X'FF'   (= 1111 1111)
0F67  C922    ST    P1+22   Output definition
                       register A (all are '1', so all
                       ports are defined as outputs)
0F69  C923    ST    P1+23   Output definition
                       register B (all are '1', so all
                       ports are defined as outputs)
0F6B  C0B2    LD    A
0F6D  C920    ST    P1 +20  write A at outputs
                       of Port A
0F6F  C0AF    LD    B
0F71  C921    ST    P1+21   write B at output
                       of Port B
(0F6B, 0F6D, 0F6F and 0F71 – the single '1' in A or B
makes one output port go high; other outputs low)
0F73  90AB    JMP to A to await next command
                       signal
```

The output remains latched until a new command signal is received. The delay time at 0F2F and 0F31 must be set to match the delay time used in the Coder program.

APPENDIX B
DATA FOR THE CONSTRUCTOR

Transistors, diodes and regulators

Diagrams of terminal connections of all types mentioned in this book appear in Figure B.1. In these diagrams the connections are shown as viewed from the *underside* of the device, except for the diodes, which are shown in side view.

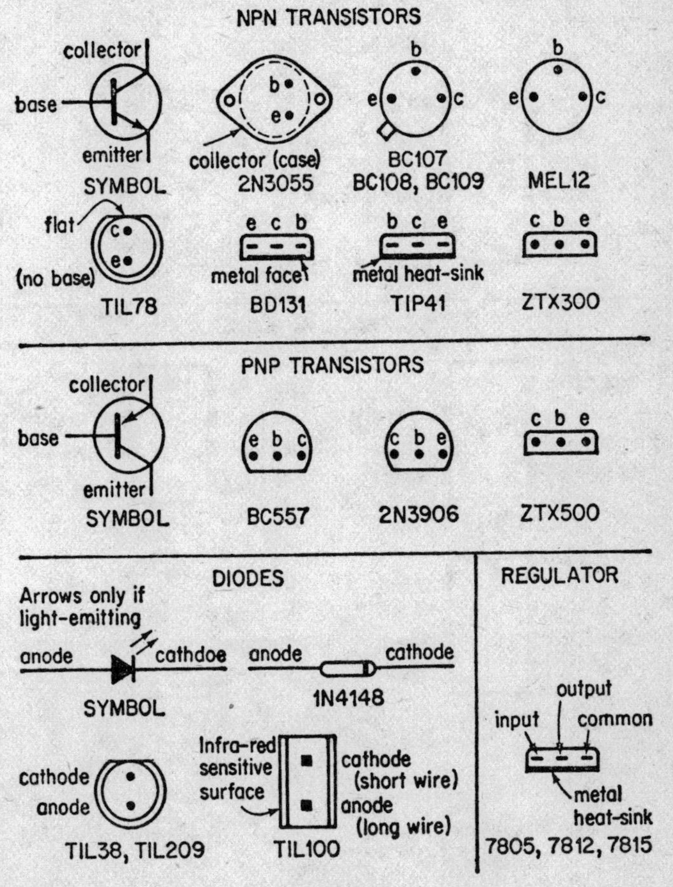

Fig. B1

Integrated circuits

Diagrams of terminal connections of the most frequently used types mentioned in this book appear in Figure B.2. In these diagrams the connections are shown as viewed *from above* when the device is mounted on a circuit-board. The connections for the following i.c.s are shown elsewhere in the book:

In Figure 10.6, p. 68: 4028, 4514, 4515, 7442, 74154
In Figure 10.8, p. 70: 74118, 74279.

MISCELLANEOUS ICs

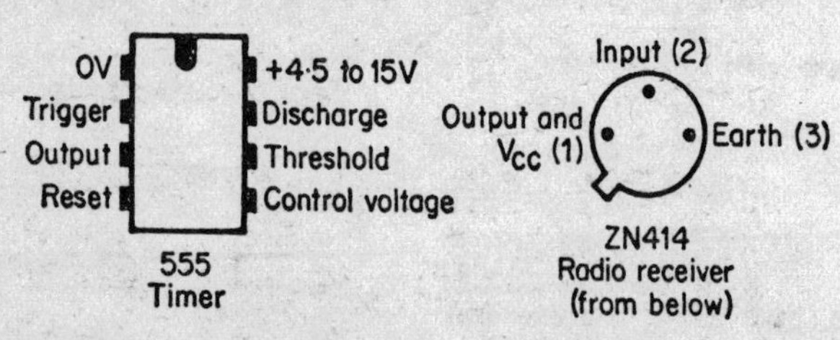

OPERATIONAL AMPLIFIERS

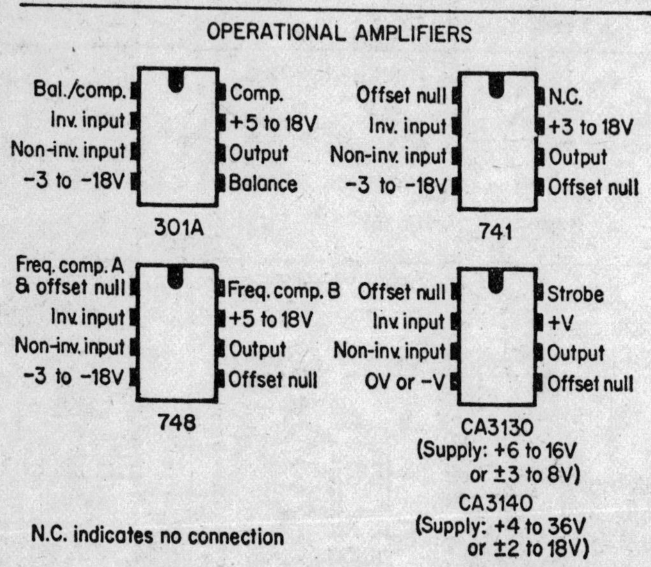

Fig. B2a

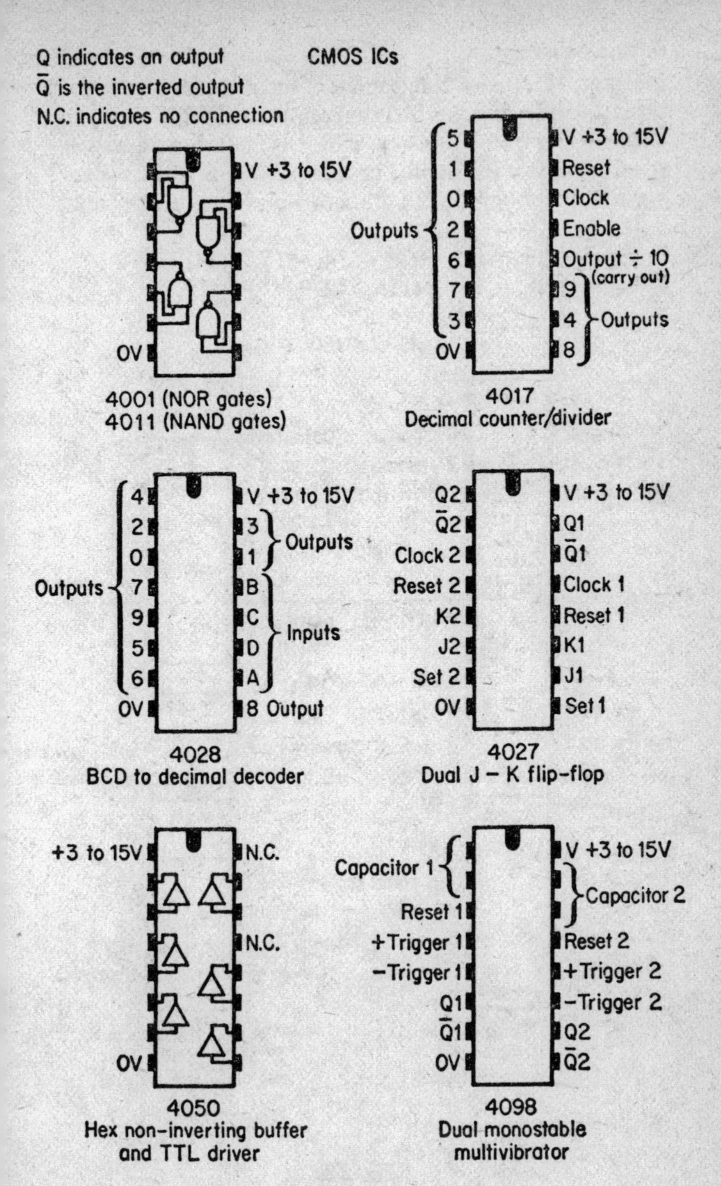
Q indicates an output
$\bar{Q}$ is the inverted output
N.C. indicates no connection
CMOS ICs
V +3 to 15V
0V
4001 (NOR gates)
4011 (NAND gates)
Outputs
5
1
0
2
6
7
3
0V
V +3 to 15V
Reset
Clock
Enable
Output ÷ 10
(carry out)
9
4
8
Outputs
4017
Decimal counter/divider
Outputs
4
2
0
7
9
5
6
0V
V +3 to 15V
3
1
Outputs
B
C
D
A
Inputs
8 Output
4028
BCD to decimal decoder
Q2
$\bar{Q}$2
Clock 2
Reset 2
K2
J2
Set 2
0V
V +3 to 15V
Q1
$\bar{Q}$1
Clock 1
Reset 1
K1
J1
Set 1
4027
Dual J – K flip-flop
+3 to 15V
N.C.
N.C.
0V
4050
Hex non-inverting buffer
and TTL driver
Capacitor 1
Reset 1
+Trigger 1
−Trigger 1
Q1
$\bar{Q}$1
0V
V +3 to 15V
Capacitor 2
Reset 2
+Trigger 2
−Trigger 2
Q2
$\bar{Q}$2
4098
Dual monostable
multivibrator

Fig. B2b

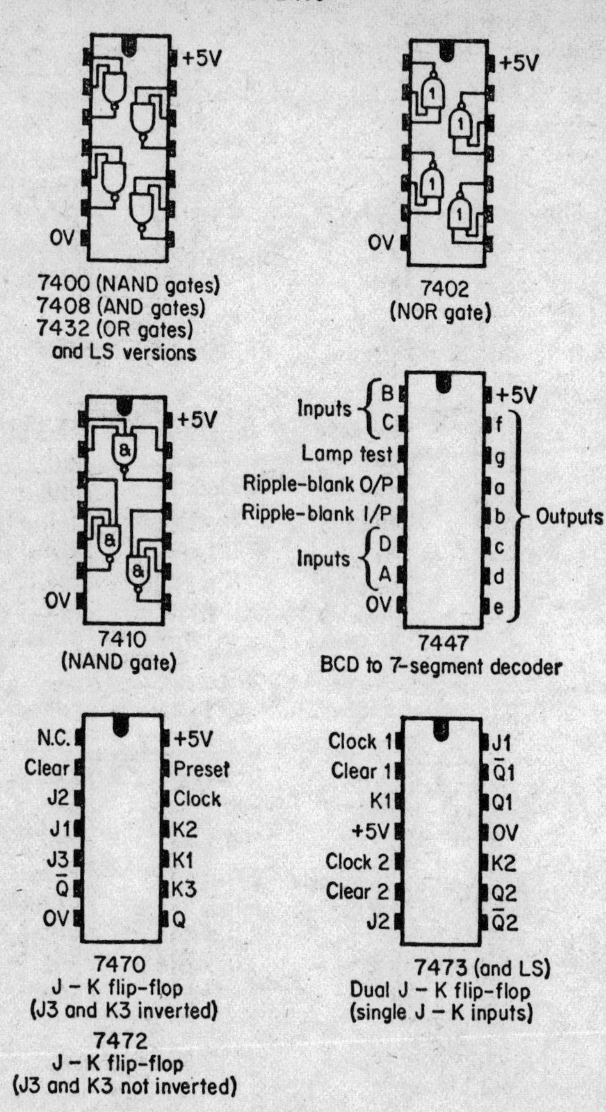
+5V
OV
7400 (NAND gates)
7408 (AND gates)
7432 (OR gates)
and LS versions
+5V
1
OV
7402
(NOR gate)
+5V
&
OV
7410
(NAND gate)
Inputs
B
C
Lamp test
Ripple-blank O/P
Ripple-blank I/P
Inputs
D
A
OV
+5V
f
g
a
b
c
d
e
Outputs
7447
BCD to 7-segment decoder
N.C.
Clear
J2
J1
J3
Q̄
OV
+5V
Preset
Clock
K2
K1
K3
Q
7470
J – K flip-flop
(J3 and K3 inverted)
7472
J – K flip-flop
(J3 and K3 not inverted)
Clock 1
Clear 1
K1
+5V
Clock 2
Clear 2
J2
J1
Q̄1
Q1
OV
K2
Q2
Q̄2
7473 (and LS)
Dual J – K flip-flop
(single J – K inputs)

Fig. B2c

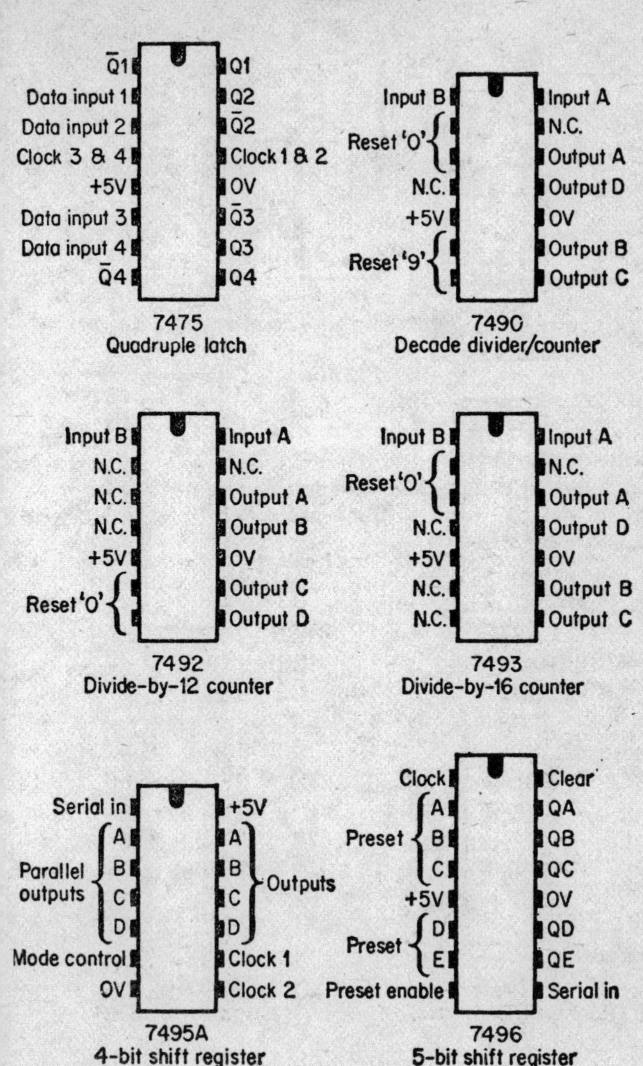
Q̄1
Data input 1
Data input 2
Clock 3 & 4
+5V
Data input 3
Data input 4
Q̄4
Q1
Q2
Q̄2
Clock 1 & 2
OV
Q̄3
Q3
Q4
7475
Quadruple latch
Input B
Reset '0'
N.C.
+5V
Reset '9'
Input A
N.C.
Output A
Output D
OV
Output B
Output C
7490
Decade divider/counter
Input B
N.C.
N.C.
N.C.
+5V
Reset '0'
Input A
N.C.
Output A
Output B
OV
Output C
Output D
7492
Divide-by-12 counter
Input B
Reset '0'
N.C.
+5V
N.C.
N.C.
Input A
N.C.
Output A
Output D
OV
Output B
Output C
7493
Divide-by-16 counter
Serial in
Parallel outputs
A
B
C
D
Mode control
OV
+5V
A
B
C
D
Outputs
Clock 1
Clock 2
7495A
4-bit shift register
Clock
Preset
A
B
C
+5V
Preset
D
E
Preset enable
Clear
QA
QB
QC
OV
QD
QE
Serial in
7496
5-bit shift register

Fig. B2d

TTL ICs (continued)

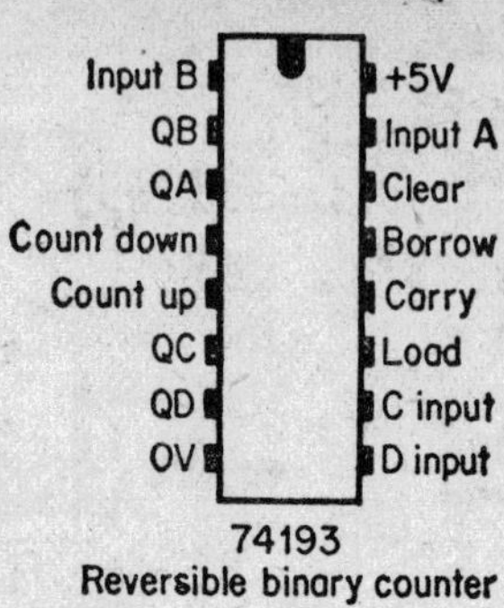

Fig B2e

Radio control circuits

The following components are required:

Transmitter:

Resistors All resistors min. res.

R17	47kΩ	R20	10Ω
R18	10kΩ	R21	3.3kΩ
R19	220Ω	R22	1.2kΩ

Capacitors

C4	0.01μF disc ceramic	C9	Trimmer 65pF
C5	3300pF ceramic	C10	100pF ceramic
C6	22pF ceramic	C11	0.01μF disc
C7	1000pF ceramic	C12	1000pF ceramic
C8	82pF ceramic		

Transistors

TR4	BC107B	TR6	BC107B
TR5	BFY52	TR7	2N3708

Crystal

XL1 Crystal of the range 26.995–27.245MHz (to match with receiver crystal). These are sold in pairs.

Miscellaneous
- RC Transmitter P.C.B.
- Circuit-board pins (pin 2145)
- Former 351 and core type 6
- Enamelled copper wire: 14 and 30 s.w.g.
- L4 Choke 33μH

Receiver:

Resistors

R1	1.5kΩ	R4	330Ω
R2	150Ω	R5	33kΩ
R3	1.5kΩ	R6	10kΩ

Capacitors

C1	10μF 25V elect. axial	C10	0.1μF polyester
C2	22pF ceramic	C11	100μF 10V elect. axial
C3	22pF ceramic	C12	0.01μF polyester
C4	0.01μF disc	C13	0.01μF polyester
C5	1000pF ceramic	C14	0.1μF polyester
C6	10μF 25V elect. axial	C15	0.1μF polyester
C7	47pF ceramic	C16	0.1μF polyester
C8	1000pF ceramic	C17	47μF 25V elect. axial
C9	0.1μF polyester		

Semiconductors

IC1	TBA651	D1	OA90

Inductors, transformers

L1	13 turns on 28 s.w.g. enam. cu wire on former 351 and core type 6
L2	4 turns on 28 s.w.g. enam. cu wire on former 351 and core type 6
L3	1.5μH choke
L4	100μH choke
T1	Toko YRCS 11098 AC2 455kHz (1st i.f.)
T2	Toko YHCS 11100 AC2 455kHz (3rd i.f.)

Miscellaneous

XL1 27MHz R/C band, selected to suit transmitter XL1

RC Receiver PCB

Pins 2145

Notes

Notes

Notes

Please note overleaf is a list of other titles that are available in our range of Radio and Electronics Books.

These should be available from most good Booksellers, Radio Component Dealers and Mail Order Companies.

However, should you experience difficulty in obtaining any title in your area, then please write directly to the publishers enclosing payment to cover the cost of the book plus adequate postage.

If you would like a copy of our latest catalogue of Radio and Electronics Books then please send a Stamped Addressed Envelope to:–

BERNARD BABANI (publishing) LTD
The Grampians
Shepherds Bush Road
London W6 7NF
England

160	Coil Design and Construction Manual	£1.95
202	Handbook of Integrated Circuits (IC's) Equivalents & Substitutes	£1.95
205	First Book of Hi-Fi Loudspeaker Enclosures	£0.95
208	Practical Stereo and Quadrophony Handbook	£0.75
214	Audio Enthusiasts Handbook	£0.85
219	Solid State Novelty Projects	£0.85
220	Build Your Own Solid State Hi-Fi and Audio Accessories	£0.85
221	28 Tested Transistor Projects	£1.25
222	Solid State Short Wave Receivers for Beginners	£1.95
223	50 Projects Using IC CA3130	£1.25
224	50 CMOS IC Projects	£1.35
225	A Practical Introduction to Digital IC's	£1.75
226	How to Build Advanced Short Wave Receivers	£1.95
227	Beginners Guide to Building Electronic Projects	£1.95
228	Essential Theory for the Electronics Hobbyist	£1.95
RCC	Resistor Colour Code Disc	£0.20
BP1	First Book of Transistor Equivalents and Substitutes	£1.50
BP2	Handbook of Radio, TV, Ind & Transmitting Tube & Valve Equivalents	£0.60
BP6	Engineers and Machinists Reference Tables	£0.75
BP7	Radio and Electronic Colour Codes and Data Chart	£0.40
BP14	Second Book of Transistor Equivalents & Substitutes	£1.75
BP24	52 Projects Using IC741	£1.75
BP27	Chart of Radio, Electronic, Semi-conductor and Logic Symbols	£0.50
BP28	Resistor Selection Handbook	£0.60
BP29	Major Solid State Audio Hi-Fi Construction Projects	£0.85
BP32	How to Build Your Own Metal and Treasure Locators	£1.95
BP33	Electronic Calculator Users Handbook	£1.50
BP34	Practical Repair and Renovation of Colour TVs	£1.95
BP36	50 Circuits Using Germanium, Silicon and Zener Diodes	£1.50
BP37	50 Projects Using Relays, SCR's and TRIACS	£1.95
BP39	50 (FET) Field Effect Transistor Projects	£1.75
BP42	50 Simple L.E.D. Circuits	£1.50
BP43	How to Make Walkie-Talkies	£1.95
BP44	IC 555 Projects	£1.95
BP45	Projects in Opto-Electronics	£1.95
BP47	Mobile Discotheque Handbook	£1.35
BP48	Electronic Projects for Beginners	£1.95
BP49	Popular Electronic Projects	£1.95
BP50	IC LM3900 Projects	£1.35
BP51	Electronic Music and Creative Tape Recording	£1.95
BP52	Long Distance Television Reception (TV-DX) for the Enthusiast	£1.95
BP53	Practical Electronics Calculations and Formulae	£2.95
BP54	Your Electronic Calculator and Your Money	£1.35
BP55	Radio Stations Guide	£1.75
BP56	Electronic Security Devices	£1.95
BP57	How to Build Your Own Solid State Oscilloscope	£1.95
BP58	50 Circuits Using 7400 Series IC's	£1.75
BP59	Second Book of CMOS IC Projects	£1.95
BP60	Practical Construction of Pre-amps, Tone Controls, Filters & Attenuators	£1.95
BP61	Beginners Guide To Digital Techniques	£0.95
BP62	Elements of Electronics – Book 1	£2.95
BP63	Elements of Electronics – Book 2	£2.25
BP64	Elements of Electronics – Book 3	£2.25
BP65	Single IC Projects	£1.50
BP66	Beginners Guide to Microprocessors and Computing	£1.95
BP67	Counter, Driver and Numeral Display Projects	£1.75
BP68	Choosing and Using Your Hi-Fi	£1.65
BP69	Electronic Games	£1.75
BP70	Transistor Radio Fault-Finding Chart	£0.50
BP71	Electronic Household Projects	£1.75
BP72	A Microprocessor Primer	£1.75
BP73	Remote Control Projects	£1.95
BP74	Electronic Music Projects	£1.75
BP75	Electronic Test Equipment Construction	£1.75
BP76	Power Supply Projects	£1.95
BP77	Elements of Electronics – Book 4	£2.95

BP78	Practical Computer Experiments	£1.75
BP79	Radio Control for Beginners	£1.75
BP80	Popular Electronic Circuits – Book 1	£1.95
BP81	Electronic Synthesiser Projects	£1.75
BP82	Electronic Projects Using Solar Cells	£1.95
BP83	VMOS Projects	£1.95
BP84	Digital IC Projects	£1.95
BP85	International Transistor Equivalents Guide	£2.95
BP86	An Introduction to BASIC Programming Techniques	£1.95
BP87	Simple L.E.D. Circuits – Book 2	£1.35
BP88	How to Use Op-Amps	£2.25
BP89	Elements of Electronics – Book 5	£2.95
BP90	Audio Projects	£1.95
BP91	An Introduction to Radio DXing	£1.95
BP92	Easy Electronics – Crystal Set Construction	£1.75
BP93	Electronic Timer Projects	£1.95
BP94	Electronic Projects for Cars and Boats	£1.95
BP95	Model Railway Projects	£1.95
BP96	C B Projects	£1.95
BP97	IC Projects for Beginners	£1.95
BP98	Popular Electronic Circuits – Book 2	£2.25
BP99	Mini-Matrix Board Projects	£1.95
BP100	An Introduction to Video	£1.95
BP101	How to Identify Unmarked IC's	£0.65
BP102	The 6809 Companion	£1.95
BP103	Multi-Circuit Board Projects	£1.95
BP104	Electronic Science Projects	£2.25
BP105	Aerial Projects	£1.95
BP106	Modern Op-Amp Projects	£1.95
BP107	30 Solderless Breadboard Projects – Book 1	£2.25
BP108	International Diode Equivalents Guide	£2.25
BP109	The Art of Programming the 1K ZX81	£1.95
BP110	How to Get Your Electronic Projects Working	£1.95
BP111	Elements of Electronics – Book 6	£3.50
BP112	A Z-80 Workshop Manual	£2.75
BP113	30 Solderless Breadboard Projects – Book 2	£2.25
BP114	The Art of Programming the 16K ZX81	£2.50
BP115	The Pre-Computer Book	£1.95
BP116	Electronic Toys Games & Puzzles	£2.25
BP117	Practical Electronic Building Blocks – Book 1	£1.95
BP118	Practical Electronic Building Blocks – Book 2	£1.95
BP119	The Art of Programming the ZX Spectrum	£2.50
BP120	Audio Amplifier Fault-Finding Chart	£0.65
BP121	How to Design and Make your Own P.C.B.s	£1.95
BP122	Audio Amplifier Construction	£2.25
BP123	A Practical Introduction to Microprocessors	£2.25
BP124	Easy Add-on Projects for Spectrum ZX81 & Ace	£2.75
BP125	25 Simple Amateur Band Aerials	£1.95
BP126	BASIC & PASCAL in Parallel	£1.50
BP127	How to Design Electronic Projects	£2.25
BP128	20 Programs for the ZX Spectrum & 16K ZX81	£1.95
BP129	An Introduction to Programming the ORIC-1	£1.95
BP130	Micro Interfacing Circuits – Book 1	£2.25
BP131	Micro Interfacing Circuits – Book 2	£2.25
BP132	25 Simple Shortwave Broadcast Band Aerials	£1.95
BP133	An Introduction to Programming the Dragon 32	£1.95
BP134	Easy Add-on Projects for Commodore 64 & Vic-20	£2.50
BP135	The Secrets of the Commodore 64	£2.50
BP136	25 Simple Indoor and Window Aerials	£1.95
BP137	BASIC & FORTRAN in Parallel	£1.95
BP138	BASIC & FORTH in Parallel	£1.95
BP139	An Introduction to Programming the BBC Model B Micro	£2.50
BP140	Digital IC Equivalents and Pin Connections	£3.95
BP141	Linear IC Equivalents and Pin Connections	£3.95
BP142	An Introduction to Programming the Acorn Electron	£1.95
BP143	An Introduction to Programming the Atari 600XL and 800XL	£2.50
BP144	Further Practical Electronics Calculations and Formulae	£3.75
BP145	25 Simple Tropical and M.W. Band Aerials	£1.95